I0047687

Cryptography Algorithms

A guide to algorithms in blockchain, quantum cryptography, zero-knowledge protocols, and homomorphic encryption

Massimo Bertaccini

Packt>

BIRMINGHAM—MUMBAI

Cryptography Algorithms

Copyright © 2022 Packt Publishing

All rights reserved. No part of this book may be reproduced, stored in a retrieval system, or transmitted in any form or by any means, without the prior written permission of the publisher, except in the case of brief quotations embedded in critical articles or reviews.

Every effort has been made in the preparation of this book to ensure the accuracy of the information presented. However, the information contained in this book is sold without warranty, either express or implied. Neither the author, nor Packt Publishing or its dealers and distributors, will be held liable for any damages caused or alleged to have been caused directly or indirectly by this book.

Packt Publishing has endeavored to provide trademark information about all of the companies and products mentioned in this book by the appropriate use of capitals. However, Packt Publishing cannot guarantee the accuracy of this information.

Group Product Manager: Vijin Boricha

Publishing Product Manager: Yogesh Deokar

Senior Editor: Arun Nadar

Content Development Editor: Sayali Pingale

Technical Editor: Nithik Cheruvakodan

Copy Editor: Safis Editing

Project Manager: Vaidehi Sawant

Proofreader: Safis Editing

Indexer: Vinayak Purushotham

Production Designer: Prashant Ghare

Marketing Co-Ordinator: Nimisha Dua

First published: January 2022

Production reference: 1190122

Published by Packt Publishing Ltd.

Livery Place

35 Livery Street

Birmingham

B3 2PB, UK.

978-1-78961-713-9

www.packt.com

*To my mom, Nadia, and my dad, Domenico, for their sacrifices
and giving me the opportunity to learn and grow.*

To my "soulmate" Elaine for her support.

– Massimo Bertaccini

Contributors

About the author

Massimo Bertaccini is a researcher and principal scientist, CEO, and co-founder of Cryptolab Inc.

His career started as a professor of mathematics and statistics. Then, he founded Cryptolab, a start-up in the field of cryptography solutions for cybersecurity. With his team of engineers, he projected and implemented the first search engine in the world that is able to work with encrypted data.

He has obtained several international prizes and awards, including the Silicon Valley inventors award and the Seal of Excellence from the EU.

Currently, he teaches mathematical models at EMUNI University as a contract professor and has published many articles in the field of cryptography, cybersecurity, and blockchain.

– At the outset, I would like to thank Yogesh Deokar (product manager) who approached me at the start and has supported me throughout the entire writing process. I would also like to extend my gratitude towards Vaidehi Sawant (project manager) and Arun Nadar (editor) for their dedication and constant support that helped me write this book smoothly from start to finish. I'd like to thank Sayali Pingale for her contribution to this book. Also, thanks to the technical reviewer, Brian Wu, for his help in validating the technical aspects. Finally, I would like to thank Alessandro Passerini, Tiziana Landi, Terenzio Carapanzano, Yi Jing Gu, Paul Hager for their constant support, and especially Gerardo Iovane, who has shared some of the most brilliant mathematical ideas with me.

About the reviewer

Xun (**Brian**) **Wu** is a senior blockchain architect and consultant. He has written eight books on popular fields within blockchain, such as Hyperledger and Ethereum (from beginning to advanced), published by O'Reilly and Packt. Brian has 18+ years of hands-on experience across various technologies, including blockchain, big data, the cloud, systems, and infrastructure UI. He has worked on more than 50 projects in his career.

Table of Contents

Section 3: New Cryptography Algorithms and Protocols

5

Introduction to Zero-Knowledge Protocols

6

New Algorithms in Public/Private Key Cryptography

7

Elliptic Curves

8

Quantum Cryptography

Section 4: Homomorphic Encryption and the Crypto Search Engine

9

Crypto Search Engine

Index

Other Books You May Enjoy

Preface

In this age of extremely high connectivity, cloud computing, ransomware, and hackers, digital assets are changing the way we live our lives, so cryptography and cybersecurity are crucial.

The changes in processing and storing data has required adequate evolution in cryptographic algorithms to advance in the eternal battle against information piracy.

This book is a good tool for students and professionals who want to focus on the next generation of cryptography algorithms.

Starting from the basics of symmetric and asymmetric algorithms, the book describes all the modern techniques of authentication, transmission, and searching on encrypted data to shelter from spies and hackers. You'll encounter Evolute algorithms with zero knowledge, consensus on the blockchain, elliptic curves, quantum cryptography, and homomorphic search.

The book even gives some focus to attacks and cryptoanalysis of the main algorithms adopted in computer science. The main purpose of this book is to train you to deeply understand this complex topic. Indeed, particular attention has been paid to present examples and suggestions are given to help you better understand the concepts.

Who this book is for

This hands-on cryptography book is for students, IT professionals, cybersecurity enthusiasts, or anyone who wants to develop their skills in modern cryptography and build a successful cybersecurity career. Working knowledge of beginner-level algebra and finite fields theory is required.

What this book covers

Chapter 1, Deep Diving into Cryptography Landscape, gives an introduction to cryptography, what it is needed for, and why it is so important in IT. This chapter also provides a panoramic view of the principal algorithms in the history of cryptography.

Chapter 2, Introduction to Symmetric Encryption, analyzes symmetric encryption. We will focus on algorithms such as DES, AES and Boolean Logic, which are widely used to implement cybersystems. Finally, we will showcase attacks to these algorithms.

Chapter 3, Asymmetric Encryption, analyzes the *classical* asymmetric encryption algorithms, such as RSA and Diffie–Hellman, and the main algorithms in private/public key encryption.

Chapter 4, Introducing Hash Functions and Digital Signatures, focuses on hash functions such as SHA-1 and looks at digital signatures, which are one of the pillars of modern cryptography. We will look at the most important and famous signatures and blind signatures (a particular case of anonymous signatures).

Chapter 5, Introduction to Zero-Knowledge Protocols, looks at zero-knowledge protocols, which are one of the new fundamental encryption protocols for the new economy of the blockchain. They are very useful for authenticating humans and machines without exposing any sensitive data in an unsafe channel of communication. New protocols, such as zk-SNARK, used in the blockchain are based on these algorithms. Finally, we will present Z/K13, a new protocol in zero knowledge, invented by the author.

Chapter 6, New Algorithms in Public/Private Key Cryptography, presents three algorithms invented by the author. MB09 is based on *Fermat's Last Theorem*. MB11 could be an alternative to RSA. Digital signatures related to these algorithms are also presented. Moreover we will present MBXX, a new protocol, invented by the author used for consensus.

Chapter 7, Elliptic Curves, looks at elliptic curves, which are the new frontier for decentralized finance. Satoshi Nakamoto adopted a particular kind of elliptic curve to implement the transmission of digital currency in Bitcoins called SECP256K1. Let's see how it works and what the main characteristics of this very robust encryption are.

Chapter 8, Quantum Cryptography, looks at how, with the advent of quantum computing, most of the algorithms we have explored until now will be under serious threat of brute-force attacks. One possible solution is **Quantum Cryptography (Q-Cryptography)**. It is one of the most exhilarating and fantastic kinds of encryption that the human mind has invented. Q-Cryptography is only at the beginning but will be widely adopted in a short time.

Chapter 9, *Crypto Search Engine*, looks at the crypto search engine, which is an outstanding application of homomorphic encryption, invented by the author. It is a search engine able to search for a query inside encrypted content. We will see how it has been implemented, the story of this enterprise, and the possible applications of this disruptive engine for security and data privacy.

To get the most out of this book

Working knowledge of beginner-level algebra and finite fields theory is required.

Download the color images

We also provide a PDF file that has color images of the screenshots/diagrams used in this book. You can download it here: `https://static.packt-cdn.com/downloads/9781789617139_ColorImages.pdf`.

Conventions used

There are a number of text conventions used throughout this book.

`Code in text`: Indicates code words in text, database table names, folder names, filenames, file extensions, pathnames, dummy URLs, mathematical equations, user input, and Twitter handles. Here is an example: "Encrypt the plaintext blocks using single DES with the `[K1]` key."

A block of code/mathematical equations are set as follows:

```
t = (3XP^2 + a)/ 2YP
t= (3*2^2 + 0)/ 2*22= 12/ 44 = Reduce[44*x == 12, x, Modulus ->
(67)] = 49
t = 49
```

Bold: Indicates a new term, an important word, or words that you see onscreen. For example, words in menus or dialog boxes appear in the text like this. Here is an example: "Hash functions are candidates for being quantum resistant, which means that hash functions can overcome a quantum computer's attack **under certain conditions**."

Tips or Important Notes
Appear like this.

Get in touch

Feedback from our readers is always welcome.

General feedback: If you have questions about any aspect of this book, mention the book title in the subject of your message and email us at customercare@packtpub.com.

Errata: Although we have taken every care to ensure the accuracy of our content, mistakes do happen. If you have found a mistake in this book, we would be grateful if you would report this to us. Please visit www.packtpub.com/support/errata, selecting your book, clicking on the Errata Submission Form link, and entering the details.

Piracy: If you come across any illegal copies of our works in any form on the Internet, we would be grateful if you would provide us with the location address or website name. Please contact us at copyright@packt.com with a link to the material.

If you are interested in becoming an author: If there is a topic that you have expertise in and you are interested in either writing or contributing to a book, please visit authors.packtpub.com.

Share Your Thoughts

Once you've read *Cryptography Algorithms*, we'd love to hear your thoughts! Scan the QR code below to go straight to the Amazon review page for this book and share your feedback.

https://packt.link/r/1789617138

Your review is important to us and the tech community and will help us make sure we're delivering excellent quality content.

Section 1:
A Brief History and
Outline of Cryptography

This section is an introductory part and aims to provide basic definitions, information, and a history of cryptography and its algorithms.

This section of the book comprises the following chapter:

- *Chapter 1, Deep Diving into Cryptography*

1
Deep Diving into Cryptography

This chapter provides an introduction to cryptography, what it is needed for, and why it is so important in IT. This chapter also gives a panoramic view of the principal algorithms from the history of cryptography, from the Caesar algorithm to the Vernam cipher and other lesser-known algorithms, such as the Beale cipher. Then, **Rivest-Shamir-Adleman (RSA)**, Diffie–Hellman, and other famous algorithms will be described in detail in the proceeding part of this book. Finally, this chapter will give you the instruments to learn cryptography and the pillars of security conservation.

In this chapter, we will cover the following topics:

- A brief introduction to cryptography
- Basic definitions and principal mathematical notations used in the book
- Binary conversion and ASCII code
- Fermat Last's Theorem, prime numbers, and modular mathematics
- The history of the principal cryptographic algorithms and an explanation of some of them (Rosetta cipher, Caesar, ROT13, Beale, Vernam)
- Security notation (semantic, provable, OTP, and so on)

An introduction to cryptography

One of the most important things in cryptography is to understand definitions and notations. I have never been a fan of definitions and notations, first of all, because I am the only one to use notations that I've invented. But I realize that it is very important, especially when we are talking about something related to mathematics, to agree among ourselves. Thus, in this section, I will introduce basic information and citations relating to cryptography.

We start with a definition of an algorithm.

In mathematics and computer science, an **algorithm** is a finite sequence of well-defined computer-implementable instructions.

An important question is: what is a cipher?

A **cipher** is a system of any type able to transform plaintext (a message) into not-intelligible text (a ciphertext or cryptogram):

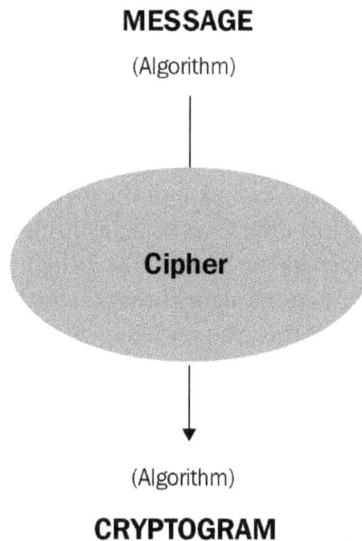

MESSAGE

(Algorithm)

Cipher

(Algorithm)

CRYPTOGRAM

Figure 1.1 – Encryption process

To get some utility from a cipher, we have to set up two operations: encryption and decryption. In simpler terms, we have to keep the message secret and safe for a certain period of time.

We define M as the set of all the messages and C as the set of all the cryptograms.

Encryption is an operation that transforms a generic message, m, into a cryptogram, c, applying a function, E:

```
m ------- > f(E) --------- > c
```

Decryption is an operation that returns the message in cleartext, m, from the cryptogram, c, applying a function, D:

```
C ------- > f(D) --------- > m
```

Mathematically, D(E(M)) = M.

This means that the E and D functions are the inverse of each other, and the E function has to be injective. **Injective** means different M values have to correspond to different C values.

Note that it doesn't matter whether I use capital letters or lowercase, such as (M) or (m); it's inconsequential at the moment. For the moment, I have used round brackets indiscriminately, but later I will use square brackets to distinguish secret elements of a function from known ones, for which I will use square brackets. So, the secret message M will be written as [M], just like any other secret parameter. Here, just showing how the algorithms work is within our scope; we'll leave their implementation to engineers.

There is another important notation that is key to encryption/decryption. To encrypt and decrypt a message, it is necessary to set up a key. In cryptography, a key is a parameter that determines the functional output of a cryptographic algorithm or cipher. Without a key, the algorithm would produce no useful results.

We define K as the set of all the keys used to encrypt and decrypt M, and k as the single encryption or decryption key, also called the session key. However, these two ways to define a key (a set of keys is K and a single key is k) will always be used, specifying what kind of key it is (private or public).

Now that we understand the main concepts of cryptographic notation, it is time to explain the difference between private and public keys:

- In cryptography, a private or secret key (Kpr), denoted as [K] or [k], is an encryption/decryption parameter known only to one, both, or multiple parties in order to exchange secret messages.
- In cryptography, a public key (Kpu) or (K) is an encryption key known by everyone who wants to send a secret message or authenticate a user.

So, what is the main difference between private and public keys?

The difference is that a private key is used both to encrypt and/or decrypt a message, while a public key is used only to encrypt a message and verify the identity (digital signatures) of humans and computers. This is a substantial and very important issue because it determines the difference between symmetric and asymmetric encryption.

Let's give a generic definition of these two methods of encryption:

- **Symmetric encryption** uses only one shared key to both encrypt and decrypt the message.

- **Asymmetric encryption** implements more parameters to generate a public key (to encrypt the message) and just one private key to decrypt the message.

As we will see later on, private keys are used in symmetric encryption to encrypt/decrypt the message with the same key and in asymmetric encryption in a general way for decryption, whereas public keys are used only in asymmetric encryption to encrypt the message and to perform digital signatures. You will see the function of these two types of keys later, but for now, keep in mind that a private key is used both in symmetric and asymmetric encryption, while a public key is used only for asymmetric encryption. Note that it's not my intention to discuss academic definitions and notation, so please try to figure out the scope and the use of each element.

One of the main problems in cryptography is the transmission of the key, or the key exchange. This problem resulted in strong diatribes in the community of mathematicians and cryptographers because it was very hard to determine how to transmit a key while avoiding physically exchanging it.

For example, if Alice and Bob wanted to exchange a key (before the advent of asymmetric encryption), the only trusted way to do that was to meet physically in one place. This condition caused a lot of problems with the massive adoption of telecommunication systems and the internet. The first problem was that internet communication relies on data exchange over unsafe channels. As you can easily understand, if Alice communicates with Bob through an insecure public communication channel, the private key has a severe possibility of being compromised, which is extremely dangerous for the security and privacy of communications.

For this reason, this question arises: *if we use a symmetric cipher to protect our secret information, how can we securely exchange the secret key?*

A simple answer is the following: we have to provide a *secure channel* of communication to exchange the key.

Someone could then reply: *how do we provide a secure channel?*

We will find the answer, or rather multiple answers, later on in this book. Even in tough military applications, such as the legendary *red line* between the leaders of the US and USSR during the Cold War, symmetric communication keys were used; nowadays, it is common to use asymmetric encryption to exchange a key. Once the key has been exchanged, the next communication session is combined with symmetric encryption to encrypt the messages transmitted.

For many reasons, asymmetric encryption is a good way to exchange a key and is good for authentication and digital signatures. Computationally, symmetric encryption is better because it can work with lower bit-length keys, saving a lot of bandwidth and timing. So, in general, its algorithms work efficiently for security using keys of 256-512 bits compared to the 4,000+ bits of asymmetric RSA encryption, for example. I will explain in detail why and how that is possible later during the analysis of the algorithms in asymmetric/ symmetric encryption.

While in this book I will analyze many kinds of cryptographic techniques, essentially, we can group all the algorithms into two big families: symmetric and asymmetric encryption.

We need some more definitions to understand cryptography well:

- **Plaintext**: In cryptography, this indicates unencrypted text, or everything that could be exposed in public. For example, (meet you tomorrow at 10 am) is plaintext.

- **Ciphertext**: In cryptography, this indicates the result of the text after having performed the encryption procedure. For example, meet you tomorrow at 10 am could become [x549559*ehebibcm3494] in ciphertext.

As I mentioned before, I use different brackets to identify plaintext and ciphertext. In particular, these brackets (...) identify plaintext, while square brackets [...] identify ciphertext.

Binary numbers, ASCII code, and notations

When we manipulate data with computers, it is common to use data as strings of 0 and 1 named bits. So, numbers can be converted into bits (base 2) rather than into base 10, like our numeric system. Let's just have a look at how the conversion mechanism works. For example, the number (123) can be written in base 10 as 1*10^2+2*10^1 +3*10^0.

Likewise, we can convert a base 10 number to a base 2 number. In this case, we use the example of the number 29:

Number 29 converted into base 2 (bit)

Step	Operation	Result	Remainder	Conversion (base 2)
Step 1:	29 / 2	14	1	**(11101)**2
Step 2:	14 / 2	7	0	
Step 3:	7 / 2	3	1	
Step 4:	3 / 2	1	1	
Step 5:	1 / 2	0	1	

Figure 1.2 – Conversion of the number 29 into base 2 (bits)

The remainder of a division is very popular in cryptography because **modular mathematics** is based on the concept of remainders. We will go deeper into this topic in the next section, when I explain prime numbers and modular mathematics.

To transform letters into a binary system to be encoded by computers, the American Standards Association invented the ASCII code in 1960.

From the ASCII website, we have the following definition:

"ASCII stands for American Standard Code for Information Interchange. It's a 7-bit character code where every single bit represents a unique character."

The following is an example of an ASCII code table with the first 10 characters:

DEC	OCT	HEX	BIN	Symbol	HTML	Number Description
0	000	00	00000000	NUL	&#000;	Null char
1	001	01	00000001	SOH	&#001;	Start of Heading
2	002	02	00000010	STX	&#002;	Start of Text
3	003	03	00000011	ETX	&#003;	End of Text
4	004	04	00000100	EOT	&#004;	End of Transmission
5	005	05	00000101	ENQ	&#005;	Enquiry
6	006	06	00000110	ACK	&#006;	Acknowledgment
7	007	07	00000111	BEL	&#007;	Bell
8	010	08	00001000	BS	&#008;	Back Space
9	011	09	00001001	HT			Horizontal Tab

10	012	0A	00001010	LF	
	Line Feed

Figure 1.3 – The first 10 characters and symbols expressed in ASCII code

Note that I will often use in my implementations, made with the **Wolfram Mathematica** research software, the character 88 as X to denote the message number to encrypt. In ASCII code, the number 88 corresponds to the symbol X, as you can see in the following example:

```
88      130     58      01011000     X      &#88;      Uppercase X
```

You can go to the *Appendix* section at the end of the book to find all the notation used in this book both for the algorithms and their implementation with Mathematica code.

Fermat's Last Theorem, prime numbers, and modular mathematics

When we talk about cryptography, we have to always keep in mind that this subject is essentially related to mathematics and logic. Before I start explaining **Fermat's Last Theorem**, I want to introduce some basic notation that will be used throughout the book to prevent confusion and for a better understanding of the topic. It's important to know that some symbols, such as =, ≡ (equivalent), and : = (this last one you can find in Mathematica to compute =), will be used by me interchangeably. It's just a way to tell you that two elements correspond to each other in equal measure; it doesn't matter whether it is in a finite field (don't worry, you will become familiar with this terminology), computer science, or in regular algebra. Mathematicians may be horrified by this, but I trust your intelligence and that you will look for the substance and not for the uniformity.

Another symbol, $\approx$ (approximate), can be used to denote similar approximative elements.

You will also encounter the $\wedge$ (exponent) symbol in cases such as in a classical way to express exponentiation: ax (a elevated to x), for example.

The $\neq$ symbol, as you should remember from high school, means **not equal** or **unequal**, which is the same as the meaning of $\not\equiv$, that is, not equivalent.

However, you will always get an explanation of the equations, so if you are not very familiar with mathematical and logical notation, you can rely on the descriptions. Anyway, I will explain each case as we come across new notation.

A prime number is an integer that can only be divided by itself and 1, for example, 2, 3, 5, 7....23....67......p.

Prime numbers are the cornerstones of mathematics because all other composite numbers originate from them.

Now, let's see what Fermat's Last Theorem is, where it is applied, and why it is useful for us.

Fermat's Last Theorem is one of the best and most beautiful theorems of classical mathematics strictly related to prime numbers. According to Wikipedia, *"In number theory, Fermat's Last Theorem (sometimes called Fermat's conjecture, especially in older texts) states that no three positive integers a, b, and c satisfy the equation $a^\wedge n + b^\wedge n = c^\wedge n$ for any integer value of n greater than 2. The cases n = 1 and n = 2 have been known since antiquity to have infinitely many solutions."*

In other words, it tells us that given the following equation, for any exponent, n>3, there is no integer, a, b, or c, that verifies the sum:

```
a^n+b^n = c^n
```

Why is this theorem so important for us? Firstly, it's because Fermat's Last Theorem is strictly related to prime numbers. In fact, given the properties of primes, in order to demonstrate Fermat's Last Theorem, it's sufficient to demonstrate the following:

```
a^p+b^p ≠ c^p
```

Here, p is any prime number greater than 2.

Fermat stated he had *a proof that was too large to fit in the margin of his notes.*

Fermat himself noted in a paper that he had a beautiful demonstration of the problem, but it has never been found.

Wiles' proof is more than 200 pages long and is immensely difficult to understand. The proof is based on elliptic curves: these curves take a particular form when they are represented in a modular form. Wiles arrived at his conclusion after 7 years and explained his proof at a mathematicians' congress in 1994. I will explain the proof and part of the logic used in *Chapter 7, Elliptic Curves*. Right now, we just assume that to demonstrate Fermat's Last Theorem, Wiles needed to rely on the Taniyama Shimura conjecture, which states that *elliptic curves over the field of rational numbers are related to modular forms*. Again, don't worry if this seems too complicated; eventually, as we progress, it will start making sense.

We will deeply analyze Fermat's Last Theorem in *Chapter 6, New Algorithms in Public/Private Key Cryptography*, when I introduce the MB09 algorithm based on Fermat's Last Theorem, among other innovative algorithms in public/private keys. Moreover, we will analyze the elliptic curves applied in cryptography in *Chapter 7, Elliptic Curves*.

Fermat was obsessed with prime numbers, just like many other mathematicians; he searched for prime numbers and their properties throughout his life. He tried to attempt to find a general formula to represent all the primes in the universe, but unluckily, Fermat, just like many other mathematicians, only managed to construct a formula for some of them. The following is *Fermat's prime numbers* formula:

```
2^2n + 1    for some positive integer n
```

If we substitute n with integers, we can obtain some prime numbers:

```
n = 1, p = 5
n = 2, p = 17
n = 3, p = 65 (not prime)
n = 4, p = 257
```

Probably more famous but very similar is the Mersenne prime numbers formula:

```
2^n - 1    for some positive integer n
n = 1, p=1
n = 2, p=3
n = 3, p=7
n = 4, p=15 (not prime)
n = 5, p=31
```

Despite countless attempts to find a formula that exclusively represents all prime numbers, nobody has reached this goal as yet.

Great Internet Mersenne Prime Search (**GIMPS**) is a research project that aims to discover the newest and biggest prime numbers with Mersenne's formula.

If you explore the GIMPS website, you can discover the following:

All exponents below 53 423 543 have been tested and verified.

All exponents below 92 111 363 have been tested at least once.

51st Known Mersenne Prime Found!

December 21, 2018 — The Great Internet Mersenne Prime Search (GIMPS) has discovered the largest known prime number, 2^82,589,933-1, having 24,862,048 digits. A computer volunteered by Patrick Laroche from Ocala, Florida made the find on December 7, 2018. The new prime number, also known as M82589933, is calculated by multiplying together 82,589,933 twos and then subtracting one. It is more than one and a half million digits larger than the previous record prime number.

Besides that, GIMPS is probably the first decentralized example of how to split CPU and computer power to reach a common goal. But why all this interest in finding big primes?

There are at least three answers to this question: the passion for pure research, the money – because there are several prizes for those who find big primes – and finally, because prime numbers are important for cryptography, just like oxygen is for humans. This is also the reason why there is prize money for discovering big prime numbers.

You will understand that most algorithms of the next generation work with prime numbers. But how do you discover whether a number is prime?

In mathematics, there is a substantial computation difference between the operation of multiplication and division. Division is a lot more computationally expensive than multiplication. This means, for instance, that if I compute 2^x, where x is a huge number, it is easy to operate the power elevation but is extremely difficult to find the divisors of that number.

Because of this, mathematicians such as Fermat struggled to find algorithms to make this computation easier.

In the field of prime numbers, Fermat produced another very interesting theorem, known as **Fermat's Last Theorem**. Before explaining this theorem, it is time to understand what modular arithmetics is and how to compute with it.

The simplest way to learn modular arithmetics is to think of a clock. When we say: *"Hey, we can meet at 1 p.m."* actually we calculate that 1 is the first hour after 12 (the clock finishes its circular *wrap*).

So, we can say that we are unconsciously calculating in modulus 12 written by the notation (mod 12), where integers *wrap around* when reaching a certain value (in this case 12), called the modulus.

Technically, the result of a calculation with a modulus consists of the remainder of the division between the number and the modulus.

For example, in our clock, we have the following:

```
13 ≡ 1 (mod 12)
```

This means that 13 is *congruent* to 1 in modulus 12. You can consider *congruent* to mean equal. In other words, we can say that the remainder of the division of 13 : 12 is 1:

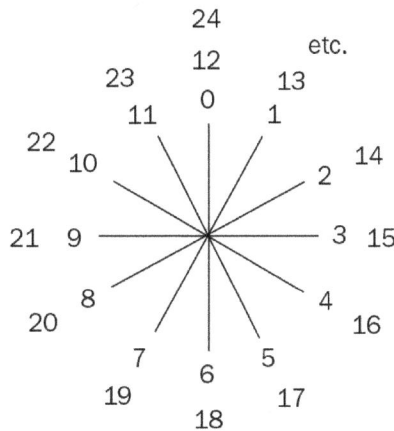

Figure 1.4 – Example of modular arithmetics with a clock

Fermat's Last Theorem states that *if* (p) *is a prime number, then for any integer* (a) *elevated to the prime number* (p) *we find* (a) *as the result of the following equation*:

```
a^p ≡ a (mod p)
```

For example, if a = 2 and p = 3, then 2^3 = 2 (mod 3). In other terms, we find the rest of the division 8 : 3 = 2 with reminder 2.

Fermat's Last Theorem is the basis of the Fermat primality test and is one of the fundamental parts of elementary number theory.

Fermat's Last Theorem states that a number, p, is *probably prime* in the following instance:

```
a^p ≡ a (mod p)
```

Now that we have refreshed our knowledge on the operations of bit conversion, we have seen what ASCII code looks like, and we have explored the basic notation of mathematics and logic, we can start our journey into cryptography.

A brief history and a panoramic overview of cryptographic algorithms

Nobody probably knows which cryptogram was the first to be invented. Cryptography has been used for a long time, approximately 4,000 years, and it has changed its paradigms a lot. First, it was a kind of hidden language, then cryptography was based on a transposition of letters in a mechanical fashion, then finally, mathematics and logic were used to solve complicated problems. What will the future hold? Probably, new methods will be invented to hide our secrets: quantum cryptography, for example, is already being experimented with and will come soon. I will explain new algorithms and methods throughout this book, but let me use this section to show you some interesting ciphers related to the *classical* period. Despite the computation power we have now, some of these algorithms have not yet been broken.

Rosetta Stone

One of the first extraordinary examples of cryptography was hieroglyphics. Cryptography means *hidden words* and comes from the union of two Greek words: κρυπτός (crypto) and γράφω (graphy). Among the many definitions of this word, we find the following: *converting ordinary plaintext into unintelligible text and vice versa*. So, we can include hieroglyphics in this definition, because we discovered how to *re-convert* their hidden meaning into intelligible text only after the *Rosetta Stone* was found. As you will probably remember from elementary school, the Rosetta Stone was written in three different languages: Ancient Egyptian (using hieroglyphics), Demotic, and Ancient Greek. The Rosetta Stone could only be decrypted because Ancient Greek was well known at the time:

Figure 1.5 – Rosetta Stone with the three languages detected

Hieroglyphics were a form of communication between the people of a country. The same problem of deciphering an unknown language could occur in the future if and when we get in contact with an alien population. A project called **SETI** (`https://www.seti.org/`) focuses on this: *"From microbes to alien intelligence, the SETI Institute is America's only organization wholly dedicated to searching for life in the universe."* Maybe if one day we get in contact with alien creatures, we will eventually understand their language. You can imagine that hieroglyphics (at the time) appeared as impenetrable as an alien language for someone who had never encountered this form of communication.

Caesar cipher

Continuing our journey through history, we find that during the Roman Empire, cryptography was used to transmit messages from the generals to the commanders and to soldiers. In fact, we find the famous **Caesar cipher**. Why is this encrypting method so famous in the history of cryptography?

This is not only because it was used by Caesar, who was one of the most valorous Roman statesmen/generals, but also because this method was probably the first that implemented mathematics.

This cipher is widely known as a shift cipher. The technique of shifting is very simple: just shift each letter you want to encrypt a fixed number of places in the alphabet so that the final effect will be to obtain a substitution of each letter for another one. So, for example, if I decide to shift by three letters, then *A* will become *D*, *E* becomes *H*, and so on.

For example, in this case, by shifting each letter three places, implicitly we have created a secret cryptographic key of [K=3]:

Caesar Cipher: Mathematical Base

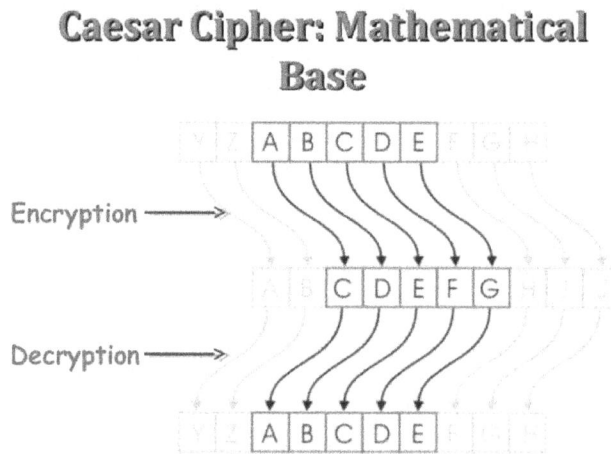

Figure 1.6 – The transposition of the letters in the Caesar cipher during the encryption and decryption processes

It is obviously a symmetric key encryption method. In this case, the algorithm works in the following way:

- Use this key: (+3).
- Message: HELLO.
- To encrypt: Take every letter and shift by +3 steps.
- To decrypt: Take every letter and de-shift by -3.

You can see in the following figure how the process of encryption and decryption of the Caesar algorithm works using key = +3; as you'll notice, the word HELLO becomes KHOOR after encryption, and then it returns to HELLO after decryption:

Key=(+3)   Key=(-3)

↓   ↓

HELLO (shift +3)   =   KHOOR (shift -3) =   HELLO

↑   ↑

Encryption Algorithm   Decryption Algorithm

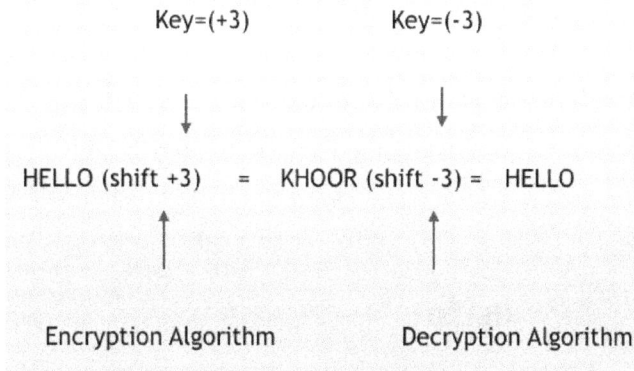

Figure 1.7 – Encryption and decryption using Caesar's algorithm

As you can imagine, the Caesar algorithm is very easy to break with a normal computer if we set a fixed key, as in the preceding example. The scheme is very simple, which, for a cryptographic algorithm, is not a problem. However, the main problem is the extreme linearity of the underlying mathematics. Using a brute-force method, that is, a test that tries all the combinations to discover the key after having guessed the algorithm used (in this case, the shift cipher), we can easily break the code. We have to check at most 25 combinations: all the letters of the English alphabet (26) minus one (that is, the same intelligible plaintext form). This is nothing compared to the billions and billions of attempts that a computer has to make in order to break a modern cryptographic algorithm.

However, there is a more complex version of this algorithm that enormously increases the efficiency of the encryption.

If I change the key for each letter and I use that key to substitute the letters and generate the ciphertext, then things become very interesting.

Let's see what happens if we encrypt HELLO using a method like this:

1. Write out the alphabet.

2. Choose a passphrase (also known as a keyphrase) such as [JULIUSCAESAR] and repeat it, putting each letter of the alphabet in correspondence with a character from the passphrase in the second row, as shown in the following screenshot.

3. After we have defined the message to encrypt, for each character composing the message (in the first row) select the corresponding character of the keyphrase (in the second row).

4. Pick up the selected corresponding characters in the second row to create the ciphertext.

Finding it a little bit complicated? Don't worry, the following example will clarify everything.

Let's encrypt HELLO with the keyphrase [JULIUSCAESARJULIUS...]:

[alphabet]	A B C D E F G H I J K L M N O P Q R S T U V W X Y Z
[passphrase]	J U L I U S C A E S A R J U L I U S C A E S A R J U
[ciphertext]	U A R L

HELLO   =   A U R R L

Figure 1.8 – Encrypting HELLO with a keyphrase becomes harder to attack

Thus, encrypting the plaintext HELLO using the alphabet and a key (or better, a passphrase or a keyphrase), JULIUSCAESAR, repeated without any spaces, we obtain the correspondent ciphertext: AURRL.

So, H becomes A, E becomes U, L becomes R (twice), and O becomes L.

Earlier, we only had to check 25 combinations to find the key in the Caesar cipher; here, things have changed a little bit, and there are 26 possibilities to discover the key! That means multiply *1*2*3...*26*, which results in *403,291,461,126,605,635,584,000,00 0*. This is a very big number. In fact, it is about one-third of all the atoms in the universe. Computationally, it is pretty hard to discover the key, even for a modern computer using a brute-force method.

Another advantage of building a cryptogram like this is that it is easy to memorize the keyword or keyphrase and hence work out the ciphertext. But let's see a cipher that is performed with a similar technique and is used in commercial contexts.

ROT13

A modern example of an algorithm that is used on the internet is **ROT13**. Essentially, this is a simple cipher derived from the Caesar cipher with a shift of (+13). Computationally, it is easy to break the Caesar cipher, but it yields an interesting effect: if we shift to the left or to the right, we will have the same result.

Just like the preceding example, in ROT13, we have to select letters that correspond to the pre-selected key. Essentially, the difference here is that instead of applying a keyphrase to perform the ciphertext, we will use 13 letters from the English alphabet as the *key generator*. ROT13 takes in encryption only the letters that occur in the English alphabet and not numbers, symbols, or other characters, which are left as they are. The ROT13 function essentially encrypts the plaintext with a key determined by the first 13 letters transposed into the second 13 letters, and the inverse for the second 13 letters.

Take a look at the following example to better understand the encryption scheme:

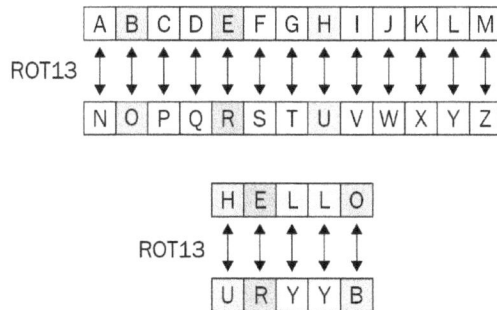

Figure 1.9 – The encryption scheme in ROT13

As you can see in the preceding diagram, H becomes U, E becomes R, L becomes Y (twice), and O becomes B:

```
HELLO = URYYB
```

The key consists of the first 13 letters of the alphabet up to M, which becomes Z, then the sequence wraps back to N, which becomes A, O becomes B, and so on to Z, which becomes M.

ROT13 was used to hide potentially offensive jokes or obscure an answer in the net.jokes newsgroup in the early 1980s.

Also, even though ROT13 is not intended to be used for a high degree of secrecy, it is still used in some cases to hide email addresses from unsophisticated spambots. ROT13 is also used for the scope of circumventing spam filters such as obscuring email content. This last function is not recommended because of the extreme vulnerability of this algorithm.

However, ROT13 was used by **Netscape Communicator** – the browser organization that released https://www.mozilla.org – to store email passwords. Moreover, ROT13 is used in Windows XP to hide some registry keys, so you can understand how sometimes even big corporations can have a lack of security and privacy in communications.

The Beale cipher

Going back to the history of cryptography, I would like to show you an amazing method of encryption whose cipher has not been decrypted yet, despite the immense computational power of our modern calculators. Very often, cryptography is used to hide precious information or fascinating treasure, just as in the mysterious story that lies behind the **Beale** cipher.

In order to better understand the method of encryption adopted in this cipher, I think it is interesting to know the story (or legend) of Beale and his treasure.

The story involves buried treasure with a value of more than $20 million, a mysterious set of encrypted documents, Wild West cowboys, and a hotel owner who dedicated his life to struggling with the decryption of these papers. The whole story is contained in a pamphlet that was published in 1885.

The story (you can find the whole version here: http://www.unmuseum.org/ bealepap.htm) begins in *January 1820 in Lynchburg, Virginia* at the *Washington Hotel* where a man named *Thomas J. Beale* checked in. The owner of the hotel, *Robert Morriss*, and *Beale* became friends, and because *Mr. Morriss* was considered a trustworthy man, he received a box containing three mysterious papers covered in numbers.

After countless troubles and many years of struggle, only the second of the three encrypted papers was deciphered.

What exactly does Beale's cipher look like?

The following content consists of three pages, containing only numbers, in an apparently random order.

The first paper is as follows:

```
71, 194, 38, 1701, 89, 76, 11, 83, 1629, 48, 94, 63, 132, 16,
111, 95, 84, 341, 975, 14, 40, 64, 27, 81, 139, 213, 63, 90,
1120, 8, 15, 3, 126, 2018, 40, 74, 758, 485, 604, 230, 436,
664, 582, 150, 251, 284, 308, 231, 124, 211, 486, 225, 401,
370, 11, 101, 305, 139, 189, 17, 33, 88, 208, 193, 145, 1, 94,
73, 416, 918, 263, 28, 500, 538, 356, 117, 136, 219, 27, 176,
130, 10, 460, 25, 485, 18, 436, 65, 84, 200, 283, 118, 320,
138, 36, 416, 280, 15, 71, 224, 961, 44, 16, 401, 39, 88, 61,
304, 12, 21, 24, 283, 134, 92, 63, 246, 486, 682, 7, 219, 184,
360, 780, 18, 64, 463, 474, 131, 160, 79, 73, 440, 95, 18, 64,
581, 34, 69, 128, 367, 460, 17, 81, 12, 103, 820, 62, 116, 97,
103, 862, 70, 60, 1317, 471, 540, 208, 121, 890, 346, 36, 150,
59, 568, 614, 13, 120, 63, 219, 812, 2160, 1780, 99, 35, 18,
21, 136, 872, 15, 28, 170, 88, 4, 30, 44, 112, 18, 147, 436,
195, 320, 37, 122, 113, 6, 140, 8, 120, 305, 42, 58, 461, 44,
106, 301, 13, 408, 680, 93, 86, 116, 530, 82, 568, 9, 102, 38,
416, 89, 71, 216, 728, 965, 818, 2, 38, 121, 195, 14, 326, 148,
234, 18, 55, 131, 234, 361, 824, 5, 81, 623, 48, 961, 19, 26,
33, 10, 1101, 365, 92, 88, 181, 275, 346, 201, 206, 86, 36,
219, 324, 829, 840, 64, 326, 19, 48, 122, 85, 216, 284, 919,
861, 326, 985, 233, 64, 68, 232, 431, 960, 50, 29, 81, 216,
```

```
321, 603, 14, 612, 81, 360, 36, 51, 62, 194, 78, 60, 200, 314,
676, 112, 4, 28, 18, 61, 136, 247, 819, 921, 1060, 464, 895,
10, 6, 66, 119, 38, 41, 49, 602, 423, 962, 302, 294, 875, 78,
14, 23, 111, 109, 62, 31, 501, 823, 216, 280, 34, 24, 150,
1000, 162, 286, 19, 21, 17, 340, 19, 242, 31, 86, 234, 140,
607, 115, 33, 191, 67, 104, 86, 52, 88, 16, 80, 121, 67, 95,
122, 216, 548, 96, 11, 201, 77, 364, 218, 65, 667, 890, 236,
154, 211, 10, 98, 34, 119, 56, 216, 119, 71, 218, 1164, 1496,
1817, 51, 39, 210, 36, 3, 19, 540, 232, 22, 141, 617, 84, 290,
80, 46, 207, 411, 150, 29, 38, 46, 172, 85, 194, 39, 261, 543,
897, 624, 18, 212, 416, 127, 931, 19, 4, 63, 96, 12, 101, 418,
16, 140, 230, 460, 538, 19, 27, 88, 612, 1431, 90, 716, 275,
74, 83, 11, 426, 89, 72, 84, 1300, 1706, 814, 221, 132, 40,
102, 34, 868, 975, 1101, 84, 16, 79, 23, 16, 81, 122, 324, 403,
912, 227, 936, 447, 55, 86, 34, 43, 212, 107, 96, 314, 264,
1065, 323, 428, 601, 203, 124, 95, 216, 814, 2906, 654, 820,
2, 301, 112, 176, 213, 71, 87, 96, 202, 35, 10, 2, 41, 17, 84,
221, 736, 820, 214, 11, 60, 760
```

The second paper (which was decrypted) is as follows:

```
115, 73, 24, 807, 37, 52, 49, 17, 31, 62, 647, 22, 7, 15, 140,
47, 29, 107, 79, 84, 56, 239, 10, 26, 811, 5, 196, 308, 85, 52,
160, 136, 59, 211, 36, 9, 46, 316, 554, 122, 106, 95, 53, 58,
2, 42, 7, 35, 122, 53, 31, 82, 77, 250, 196, 56, 96, 118, 71,
140, 287, 28, 353, 37, 1005, 65, 147, 807, 24, 3, 8, 12, 47,
43, 59, 807, 45, 316, 101, 41, 78, 154, 1005, 122, 138, 191,
16, 77, 49, 102, 57, 72, 34, 73, 85, 35, 371, 59, 196, 81, 92,
191, 106, 273, 60, 394, 620, 270, 220, 106, 388, 287, 63, 3,
6, 191, 122, 43, 234, 400, 106, 290, 314, 47, 48, 81, 96, 26,
115, 92, 158, 191, 110, 77, 85, 197, 46, 10, 113, 140, 353,
48, 120, 106, 2, 607, 61, 420, 811, 29, 125, 14, 20, 37, 105,
28, 248, 16, 159, 7, 35, 19, 301, 125, 110, 486, 287, 98, 117,
511, 62, 51, 220, 37, 113, 140, 807, 138, 540, 8, 44, 287, 388,
117, 18, 79, 344, 34, 20, 59, 511, 548, 107, 603, 220, 7, 66,
154, 41, 20, 50, 6, 575, 122, 154, 248, 110, 61, 52, 33, 30,
5, 38, 8, 14, 84, 57, 540, 217, 115, 71, 29, 84, 63, 43, 131,
29, 138, 47, 73, 239, 540, 52, 53, 79, 118, 51, 44, 63, 196,
12, 239, 112, 3, 49, 79, 353, 105, 56, 371, 557, 211, 505, 125,
360, 133, 143, 101, 15, 284, 540, 252, 14, 205, 140, 344, 26,
811, 138, 115, 48, 73, 34, 205, 316, 607, 63, 220, 7, 52, 150,
44, 52, 16, 40, 37, 158, 807, 37, 121, 12, 95, 10, 15, 35, 12,
131, 62, 115, 102, 807, 49, 53, 135, 138, 30, 31, 62, 67, 41,
85, 63, 10, 106, 807, 138, 8, 113, 20, 32, 33, 37, 353, 287,
140, 47, 85, 50, 37, 49, 47, 64, 6, 7, 71, 33, 4, 43, 47, 63,
```

```
1, 27, 600, 208, 230, 15, 191, 246, 85, 94, 511, 2, 270, 20,
39, 7, 33, 44, 22, 40, 7, 10, 3, 811, 106, 44, 486, 230, 353,
211, 200, 31, 10, 38, 140, 297, 61, 603, 320, 302, 666, 287,
2, 44, 33, 32, 511, 548, 10, 6, 250, 557, 246, 53, 37, 52, 83,
47, 320, 38, 33, 807, 7, 44, 30, 31, 250, 10, 15, 35, 106, 160,
113, 31, 102, 406, 230, 540, 320, 29, 66, 33, 101, 807, 138,
301, 316, 353, 320, 220, 37, 52, 28, 540, 320, 33, 8, 48, 107,
50, 811, 7, 2, 113, 73, 16, 125, 11, 110, 67, 102, 807, 33, 59,
81, 158, 38, 43, 581, 138, 19, 85, 400, 38, 43, 77, 14, 27, 8,
47, 138, 63, 140, 44, 35, 22, 177, 106, 250, 314, 217, 2, 10,
7, 1005, 4, 20, 25, 44, 48, 7, 26, 46, 110, 230, 807, 191, 34,
112, 147, 44, 110, 121, 125, 96, 41, 51, 50, 140, 56, 47, 152,
540, 63, 807, 28, 42, 250, 138, 582, 98, 643, 32, 107, 140,
112, 26, 85, 138, 540, 53, 20, 125, 371, 38, 36, 10, 52, 118,
136, 102, 420, 150, 112, 71, 14, 20, 7, 24, 18, 12, 807, 37,
67, 110, 62, 33, 21, 95, 220, 511, 102, 811, 30, 83, 84, 305,
620, 15, 2, 10, 8, 220, 106, 353, 105, 106, 60, 275, 72, 8,
50, 205, 185, 112, 125, 540, 65, 106, 807, 138, 96, 110, 16,
73, 33, 807, 150, 409, 400, 50, 154, 285, 96, 106, 316, 270,
205, 101, 811, 400, 8, 44, 37, 52, 40, 241, 34, 205, 38, 16,
46, 47, 85, 24, 44, 15, 64, 73, 138, 807, 85, 78, 110, 33, 420,
505, 53, 37, 38, 22, 31, 10, 110, 106, 101, 140, 15, 38, 3, 5,
44, 7, 98, 287, 135, 150, 96, 33, 84, 125, 807, 191, 96, 511,
118, 40, 370, 643, 466, 106, 41, 107, 603, 220, 275, 30, 150,
105, 49, 53, 287, 250, 208, 134, 7, 53, 12, 47, 85, 63, 138,
110, 21, 112, 140, 485, 486, 505, 14, 73, 84, 575, 1005, 150,
200, 16, 42, 5, 4, 25, 42, 8, 16, 811, 125, 160, 32, 205, 603,
807, 81, 96, 405, 41, 600, 136, 14, 20, 28, 26, 353, 302, 246,
8, 131, 160, 140, 84, 440, 42, 16, 811, 40, 67, 101, 102, 194,
138, 205, 51, 63, 241, 540, 122, 8, 10, 63, 140, 47, 48, 140,
288
```

The third paper is as follows:

```
317, 8, 92, 73, 112, 89, 67, 318, 28, 96,107, 41, 631, 78, 146,
397, 118, 98, 114, 246, 348, 116, 74, 88, 12, 65, 32, 14, 81,
19, 76, 121, 216, 85, 33, 66, 15, 108, 68, 77, 43, 24, 122, 96,
117, 36, 211, 301, 15, 44, 11, 46, 89, 18, 136, 68, 317, 28,
90, 82, 304, 71, 43, 221, 198, 176, 310, 319, 81, 99, 264, 380,
56, 37, 319, 2, 44, 53, 28, 44, 75, 98, 102, 37, 85, 107, 117,
64, 88, 136, 48, 151, 99, 175, 89, 315, 326, 78, 96, 214, 218,
311, 43, 89, 51, 90, 75, 128, 96, 33, 28, 103, 84, 65, 26, 41,
246, 84, 270, 98, 116, 32, 59, 74, 66, 69, 240, 15, 8, 121, 20,
77, 89, 31, 11, 106, 81, 191, 224, 328, 18, 75, 52, 82, 117,
201, 39, 23, 217, 27, 21, 84, 35, 54, 109, 128, 49, 77, 88, 1,
```

```
81, 217, 64, 55, 83, 116, 251, 269, 311, 96, 54, 32, 120, 18,
132, 102, 219, 211, 84, 150, 219, 275, 312, 64, 10, 106, 87,
75, 47, 21, 29, 37, 81, 44, 18, 126, 115, 132, 160, 181, 203,
76, 81, 299, 314, 337, 351, 96, 11, 28, 97, 318, 238, 106, 24,
93, 3, 19, 17, 26, 60, 73, 88, 14, 126, 138, 234, 286, 297,
321, 365, 264, 19, 22, 84, 56, 107, 98, 123, 111, 214, 136, 7,
33, 45, 40, 13, 28, 46, 42, 107, 196, 227, 344, 198, 203, 247,
116, 19, 8, 212, 230, 31, 6, 328, 65, 48, 52, 59, 41, 122, 33,
117, 11, 18, 25, 71, 36, 45, 83, 76, 89, 92, 31, 65, 70, 83,
96, 27, 33, 44, 50, 61, 24, 112, 136, 149, 176, 180, 194, 143,
171, 205, 296, 87, 12, 44, 51, 89, 98, 34, 41, 208, 173, 66, 9,
35, 16, 95, 8, 113, 175, 90, 56, 203, 19, 177, 183, 206, 157,
200, 218, 260, 291, 305, 618, 951, 320, 18, 124, 78, 65, 19,
32, 124, 48, 53, 57, 84, 96, 207, 244, 66, 82, 119, 71, 11,
86, 77, 213, 54, 82, 316, 245, 303, 86, 97, 106, 212, 18, 37,
15, 81, 89, 16, 7, 81, 39, 96, 14, 43, 216, 118, 29, 55, 109,
136, 172, 213, 64, 8, 227, 304, 611, 221, 364, 819, 375, 128,
296, 1, 18, 53, 76, 10, 15, 23, 19, 71, 84, 120, 134, 66, 73,
89, 96, 230, 48, 77, 26, 101, 127, 936, 218, 439, 178, 171, 61,
226, 313, 215, 102, 18, 167, 262, 114, 218, 66, 59, 48, 27, 19,
13, 82, 48, 162, 119, 34, 127, 139, 34, 128, 129, 74, 63, 120,
11, 54, 61, 73, 92, 180, 66, 75, 101, 124, 265, 89, 96, 126,
274, 896, 917, 434, 461, 235, 890, 312, 413, 328, 381, 96, 105,
217, 66, 118, 22, 77, 64, 42, 12, 7, 55, 24, 83, 67, 97, 109,
121, 135, 181, 203, 219, 228, 256, 21, 34, 77, 319, 374, 382,
675, 684, 717, 864, 203, 4, 18, 92, 16, 63, 82, 22, 46, 55, 69,
74, 112, 134, 186, 175, 119, 213, 416, 312, 343, 264, 119, 186,
218, 343, 417, 845, 951, 124, 209, 49, 617, 856, 924, 936, 72,
19, 28, 11, 35, 42, 40, 66, 85, 94, 112, 65, 82, 115, 119, 236,
244, 186, 172, 112, 85, 6, 56, 38, 44, 85, 72, 32, 47, 63, 96,
124, 217, 314, 319, 221, 644, 817, 821, 934, 922, 416, 975, 10,
22, 18, 46, 137, 181, 101, 39, 86, 103, 116, 138, 164, 212,
218, 296, 815, 380, 412, 460, 495, 675, 820, 952
```

The second cipher was successfully decrypted around 1885. Here, I will discuss the main considerations about this kind of cipher.

Since the numbers in the cipher far exceed the number of letters in the alphabet, we can assume that it is not a substitution nor a transposition cipher. So, we can assume that each number represents a letter, but this letter is obtained from a word contained in an external text. A cipher following this criterion is called a **book cipher**: in the case of a book cipher, a book or any other text could be used as a key. Now, the effective key here is the method of obtaining the letters from the text.

Using this system, the second cipher was decrypted by drawing on the United States Declaration of Independence. Assigning a number to each word of the referring text (the United States Declaration of Independence) and picking up the first letter of each word selected in the key (the list of the numbers, in this case, referred to the second cipher), we can extrapolate the plaintext. The extremely intelligent trick of this cipher is that the key text (the United States Declaration of Independence) is public but at the same time it was unknown to the entire world except for whom the message was intended. Only when someone holds the key (the list of the numbers) and the "key text" can they easily decrypt the message.

Let's look at the process of decrypting the second cipher:

1. Assign to each word of the text a number in order from the first to the last word.

2. Extrapolate the first letter of each word using the numbers contained in the cipher.

3. Read the plaintext.

The following is the first part of the United States Declaration of Independence (until the 115th word) showing each word with its corresponding number:

```
When(1)  in(2)  the(3)  course(4)  of(5)  human(6)  events(7)
it(8)  becomes(9)  necessary(10)  for(11)  one(12)  people(13)
to(14)  dissolve(15)  the(16)  political(17)  bands(18)  which(19)
have(20)  connected(21)  them(22)  with(23)  another(24)  and(25)
to(26)  assume(27)  among(28)  the(29)  powers(30)  of(31)  the(32)
earth(33)  the(34)  separate(35)  and(36)  equal(37)  station(38)
to(39)  which(40)  the(41)  laws(42)  of(43)  nature(44)  and(45)
of(46)  nature's(47)  god(48)  entitle(49)  them(50)  a(51)
decent(52)  respect(53)  to(54)  the(55)  opinions(56)  of(57)
mankind(58)  requires(59)  that(60)  they(61)  should(62)
declare(63)  the(64)  causes(65)  which(66)  impel(67)  them(68)
to(69)  the(70)  separation(71)  we(72)  hold(73)  these(74)
truths(75)  to(76)  be(77)  self(78)  evident(79)  that(80)  all(81)
men(82)  are(83)  created(84)  equal(85)  that(86)  they(87)  are(88)
endowed(89)  by(90)  their(91)  creator(92)  with(93)  certain(94)
unalienable(95)  rights(96)  that(97)  among(98)  these(99)
are(100)  life(101)  liberty(102)  and(103)  the(104)  pursuit(105)
of(106)  happiness(107)  that(108)  to(109)  secure(110)  these(111)
rights(112)  governments(113)  are(114)  instituted(115) ...
```

The following numbers represent the first rows of the second cipher; as you can see, the bold words (with their corresponding numbers) correspond to the numbers we find in the ciphertext:

```
115, 73, 24, 807, 37, 52, 49, 17, 31, 62, 647, 22, 7, 15, 140,
47, 29, 107, 79, 84, 56, 239, 10, 26, 811, 5, 196, 308, 85, 52,
160, 136, 59, 211, 36, 9, 46, 316, 554, 122, 106, 95, 53, 58,
2, 42, 7, 35...
```

The following is the result of decryption using the cipher combined with the key text (the United States Declaration of Independence), picking up the first letter of each corresponding word, that is, the plaintext:

```
115 = instituted = I
73 = hold = h
24 = another = a
807 (missing) = v
37 = equal = e
52 = decent = d
49 = entitle = e
```

And so on…

I haven't included the entire United States Declaration of Independence; these are only the first 115 words. But if you want, you can visit http://www.unmuseum.org/bealepap.htm and try the exercise to rebuild the entire plaintext.

Here (with some missing letters) is the reconstruction of the first sentence:

```
I have deposited in the county of Bedford…….
```

If we carry on and compare the numbers with the corresponding numbers of the initial letters of the United States Declaration of Independence, the decryption will be as follows:

I have deposited in the county of Bedford, about four miles from Buford's, in an excavation or vault, six feet below the surface of the ground, the following articles, belonging jointly to the parties whose names are given in number "3," herewith:

The first deposit consisted of one thousand and fourteen pounds of gold, and three thousand eight hundred and twelve pounds of silver, deposited November, 1819. The second was made December, 1821, and consisted of nineteen hundred and seven pounds of gold, and twelve hundred and eighty-eight pounds of silver; also jewels, obtained in St. Louis in exchange for silver to save transportation, and valued at $13,000.

The above is securely packed in iron pots, with iron covers. The vault is roughly lined with stone, and the vessels rest on solid stone, and are covered with others. Paper number "1" describes the exact locality of the vault so that no difficulty will be had in finding it.

Many other cryptographers and cryptologists have tried to decrypt the first and third Beale ciphers in vain. Others, such as the treasure hunter *Mel Fisher*, who discovered hundreds of millions of dollars' worth of valuables under the sea, went to Bedford to search the area in order to find the treasure, without success.

Maybe Beale's tale is just a legend. Or maybe it is true, but nobody will ever know where the treasure is because nobody will decrypt the first cipher. Or, the treasure will never be unearthed because someone has already found it.

Anyway, what is really interesting in this story is the implementation of such a strong cipher without the help of any computers or electronic machines; it was just made with brainpower, a pen, and a sheet of paper.

Paradoxically, the number of attempts required to crack the cipher go from 1 to infinity assuming that the attacker works with brute force, exploring all the texts written in the world at that moment. On top of that, what happens if a key text is not public but was written by the transmitter himself and has been kept secret? In this case, if the cryptologist doesn't have the key (so doesn't hold the key text), the likelihood of them decrypting the cipher is *zero*.

The Beale cipher is also interesting because this kind of algorithm could have new applications in modern cryptography or in the future. Some of these applications could be related to methods of research for encrypted data in cloud computing.

The Vernam cipher

The **Vernam cipher** has the highest degree of security for a cipher, as it is theoretically completely secure. Since it uses a truly random key of the same length as the plaintext, it is called the **perfect cipher**. It's just a matter of entropy and randomness based on Shannon's principle of information entropy that determines an equal probability of each bit contained in the ciphertext. We will revisit this algorithm in *Chapter 8, Quantum Cryptography*, where we talk about quantum key distribution and the related method to encrypt the plaintext after determining the quantum key. Another interesting implementation is Hyper Crypto Satellite, which uses this algorithm to encrypt the plaintext crafted by a random key transmitted by a satellite radiocommunication and expressed as an infinite string of bits.

But for now, let's go on to explore the main characteristics of this algorithm.

The essential element of the algorithm is using the key only once per session. This feature makes the algorithm invulnerable to attacks against the ciphertext and even in the unlikely event that the key is stolen, it would be changed at the time of the next transmission.

The method is very simple: by adding the key to the message (mod 2) bit by bit, we will obtain the ciphertext. We will see this method, called XOR, many times throughout this book, especially when we discuss symmetric encryption in *Chapter 2*, *Introduction to Symmetric Encryption*. Just remember that the key has to be of the same length as the message.

A numerical example is as follows:

- 00101001 (plaintext)
- 10101100 (key): Adding each bit (mod 2)
- 10000101 (ciphertext)

Step 1: Transform the plaintext into a string of bits using ASCII code.

Step 2: Generate a random key of the same length as the plaintext.

Step 3: Encrypt by adding modulo 2 (XOR) of the plaintext bitwise to the key and obtain the ciphertext.

Step 4: Decrypt by making the inverse operation of adding the ciphertext to the key and obtain the plaintext again.

To make an example with numbers and letters, we will go back to HELLO. Let's assume that each letter corresponds to a number, starting from 0 = A, 1 = B, 2 = C, 3 = D, 4 = E ... and so on until 25 = Z.

The random key is [DGHBC].

The encryption will present the following transposition:

Plaintext		H	E	L	L	O
		7	4	11	11	14
Key	=	D	G	H	B	C
	+	3	6	7	1	2
	=	10	10	18	12	16
Ciphertext		K	K	S	M	Q

Figure 1.10 – Encryption scheme in the Vernam algorithm

So, after transposing the letters, the encryption of [HELLO] is [KKSMQ].

You can create an exercise by yourself to decrypt the [KKSMQ] ciphertext using the inverse process: applying f(-K) to the ciphertext, returning the HELLO plaintext.

I would just like to remark that this algorithm is very strong if well implemented, following all the warnings and instructions to avoid a drastic reduction of security. One of the attacks that many algorithms suffer is well known as a ciphertext-only attack. This is successful if the attacker can deduce the plaintext, or even better the key, using the ciphertext or pieces of it. The most common techniques are frequency analysis and traffic analysis.

This algorithm is not vulnerable to ciphertext-only attacks. Moreover, if a piece of a key is known, it will be possible to decipher only the piece corresponding to the related bits. The rest of the ciphertext will be difficult to decrypt if it is long enough. However, the conditions regarding the implementation of this algorithm are very restrictive in order to obtain absolute invulnerability. First of all, the generation of the key has to be completely random. Second, the key and the message have to be of the same length, and third, there is always the problem of the key transmission.

This last problem affects all symmetric algorithms and is basically the problem that pushed cryptographers to invent asymmetric encryption to exchange keys between Alice and Bob (which we will see in the next chapter).

The second problem concerns the length of the key: if the message is too short, for instance, the word ten, to indicate the time of a military attack, the attacker could also rely on their good sense or on luck. It doesn't matter if there is a random key for a short message. The message could be decrypted intuitively if the attacker knows the topic of the transmission. On the other hand, if the message is very long, we are forced to use a very long key. In this case, the key will be very expensive to produce and expensive to transmit. Moreover, considering that for every new transmission the key has to be changed, the cost of implementing this cipher for commercial purposes is very high.

This is why, in general, *mono-use strings* such as this were used for military purposes during the Second World War and after. As I said before, this was the legendary algorithm used for the *red line* between Washington and Moscow to encrypt communications between the leaders of the US and the USSR during the Cold War.

Finally, we will analyze the implementation of this algorithm. It could be difficult to find a way to generate and transmit a random key, even if the security of the method is very high. In the last section of this book, I will show a new method for the transmission and the implementation of keys using the Vernam cipher combined with other algorithms and methods. This new **one-time pad** (**OTP**) system, named *Hyper Crypto Satellite*, could be used for both the authentication and the encryption of messages. I will also show you the possible vulnerabilities of the system and how to generate a very random key. The method was a candidate at the **Satellite International Conference on Space**, but at the time I decided not to present it to the public.

Notes on security and computation

All the algorithms we have seen in this chapter are symmetric. The basic problem that remains unsolved is the transmission of the key. As I've already said, this problem will be overcome by the asymmetric cryptography that we will explore in the next chapter. In this section, we will analyze the computational problem related to the security of cryptographic algorithms generally speaking. Later in the book, we will focus on the security of any algorithm we will analyze.

To make a similitude, we can say that in cryptography the weak link of the chain destroys the entire chain. That is the same problem as using a very strong cryptographic algorithm to protect the data but leaving its password on the computer screen. In other words, a cryptographic algorithm has to be made of a similar grade of security with respect to mathematical problems. To clarify this concept with an example: factorization and discrete logarithm problems hold similar computational characteristics for now; however, if tomorrow one of these problems were solved, then an algorithm that is based on both would be unuseful.

Let's go deeper to analyze some of the principles universally recognized in cryptography. The first statement is *Cryptography has to be open source.*

With the term *open source*, I am referring to the algorithm and not, obviously, to the key. In other words, we have to rely on **Kerckhoffs' principle**, which states the following:

"A cryptosystem should be secure even if everything about the system, except the key, is public knowledge.

Kerckhoffs' principle applies beyond codes and ciphers to security systems in general: every secret creates a potential failure point. Secrecy, in other words, is a prime cause of brittleness—and therefore something likely to make a system prone to catastrophic collapse. Conversely, openness provides ductility."

– Bruce Schneier

In practice, the algorithm that underlies the encryption code has to be known. It's not useful and is also dangerous to rely on the secrecy of the algorithm in order to exchange secret messages. The reason is that, essentially, if an algorithm has to be used by an open community (just like the internet), it is impossible to keep it secret.

The second statement is *The security of an algorithm depends largely on its underlying mathematical problem.*

As an example, RSA, one of the most famous and most widely used algorithms in the history of cryptography, is supported by the mathematical problem of factorization.

Factorization is essentially the decomposition of a number into its divisors:

```
21 = 3 x 7
```

It's very easy to find the divisors of 21, which are 3 and 7, for small integers, but it is also well known that increasing the number of digits will exponentially increase the problem of factorization.

We will deeply analyze asymmetric algorithms such as RSA in this book, and in particular, in *Chapter 3*, *Asymmetric Encryption*, when I will explain asymmetric encryption. But here, it is sufficient to explain why RSA is used to protect financial, intelligence, and other kinds of very sensitive secrets.

The reason is that the mathematical problem underlining RSA (factorization) is still a hard problem to solve for computers of this generation. However, in this introductory section, I can't go deeper into analyzing RSA, so I will limit myself to saying that RSA suffers from not only the problem of factorization as its point of attack, but there is another equally competitive, in computational terms, problem, which is the **discrete logarithm** problem. Later in the book, we will even analyze both these hard computational problems. Now, we assume (incorrectly, as 99% of cryptographic texts do) that the pillar of security underlying RSA is factorization. In *Chapter 6*, *New Algorithms in Public/Private Key Cryptography*, I will show an attack on the RSA algorithm depending on a problem different from factorization. It's the similitude of the weak link of the chain explained at the beginning of this section. If something in an algorithm goes wrong, the underlying security of the algorithm fails.

Anyway, let's see what happens when we attempt to break RSA relying only on the factorization problem, using brute force. In this case, just to give you an idea of the computational power required to decompose an RSA number of 250 digits, factorizing a big semi-prime number is not easy at all if we are dealing with hundreds of digits, or thousands. Just to give you a demonstration, RSA-250 is an 829-bit number composed of 250 decimal digits and is very hard to break with a computer from the current generation.

This integer was factorized in *February 2020*, with a total computation time of about 2,700 core years with **Intel Xeon Gold 6130** at 2.1 GHz. Like many factorization records, this one was performed using a grid of 100 machines and an optimization algorithm that elevated their computation.

The third statement is *Practical security is always less secure than theoretical security*.

For example, if we analyze the Vernam cipher, we can easily understand how the implementation of this algorithm in practice is very difficult. So, we can say that Vernam is invulnerable but only in theoretical security, not in practical security. A corollary of this assumption is this: implementing an algorithm means putting into practice its theoretical scheme and adding much more complexity to it. So, *complexity is the enemy of security*. The more complex a system is, the more points of attack can be found.

Another consideration is related to the grade of security of an algorithm. We can better understand this concept by considering Shannon's theory and the concept of *perfect secrecy*. The definition given by Claude Shannon in 1949 of perfect secrecy is based on statistics and probabilities. However, about the maximum grade of security, Shannon theorized that a ciphertext maintains perfect secrecy if an attacker's knowledge of the content of a message is the same both before and after the adversary inspects the ciphertext, attacking it with unlimited resources. That is, the message gives the adversary precisely no information about the message contents.

To better understand this concept, I invite you to think of different levels or grades of security, in which any of these degrees is secure but with a decreasing gradient. In other words, the highest level is the strongest and the lowest is the weakest but in the middle, there is a zone of an indefinite grade that depends on the technological computational level of the adversary.

It's not important how many degrees are supposed to be secure and how many are not. I think that, essentially, we have to consider what is certainly secure and what is not, but also what can be accepted as secure in a determinate time. With that in mind, let's see the difference between perfect secrecy and secure:

- A cryptosystem could be considered to have *perfect secrecy* if it satisfies at least two conditions:

  - It cannot be broken even if the adversary has unlimited computing power.

  - It's not possible to get any information about the message, [m], and the key, [k], by analyzing the cryptogram, [c] (that is, Vernam is a theoretically perfect secrecy system but only under determinate conditions).

- A cryptogram is *secure* even if, theoretically, an adversary can break the cryptosystem (that is, if they had quantum computational power and an algorithm of factorization that runs well) but the underlying mathematical problem is considered at that time very hard to solve. Under some conditions, ciphers can be used (such as RSA, Diffie-Hellmann, and El Gamal) because, based on empirical evidence, factorization and discrete logarithms are still hard problems to solve.

So, the concept of security is dynamic and very fuzzy. What is secure now might not be tomorrow. What will happen to RSA and all of the classical cryptography if quantum computers become effective, or a powerful algorithm is discovered tomorrow that is able to break the factorization problem? We will come back to these questions in *Chapter 8, Quantum Cryptography*. For now, I can say that most *classical* cryptography algorithms will be broken by the disruptive computational power of quantum computers, but we don't know yet when this will happen.

Under some conditions, we will see that the **quantum exchange of the key** can be considered a **perfect secrecy system**. But it doesn't always work, so it's not currently used. Some OTP systems could now be considered highly secure (maybe semi-perfect secrecy), but everything depends on the practical implementation. Finally, remember an important rule: a weak link in the chain destroys everything.

So, in conclusion, we can note the following:

- Cryptography has to be open source (the algorithms have to be known), except for the key.
- The security of an algorithm depends largely on its underlying mathematical problem.
- Complexity is the enemy of security.
- Security is a dynamic concept: perfect security is only a theoretical issue.

Summary

In this chapter, we have covered the basic definitions of cryptography; we have refreshed our knowledge of the binary system and ASCII code, and we also explored prime numbers, Fermat's equations, and modular mathematics. Then, we had an overview of classical cryptographic algorithms such as Caesar, Beale, and Vernam.

Finally, in the last section, we analyzed security in a philosophical and technical way, distinguishing the grade of security in cryptography in relation to the grade of complexity.

In the next chapter, we will explore symmetric encryption, where we deep dive into algorithms such as the **Data Encryption Standard** (**DES**) and **Advanced Encryption Standard** (**AES**) families, and also address some of the issues mentioned in this chapter.

Section 2:
Classical Cryptography (Symmetric and Asymmetric Encryption)

This section will deeply analyze classical cryptography: symmetric and asymmetric encryption, hash functions, and digital signatures. It will guide you through the most famous algorithms used in cybersecurity and ICT.

This section of the book comprises the following chapters:

- *Chapter 2, Introduction to Symmetric Encryption*
- *Chapter 3, Asymmetric Encryption*
- *Chapter 4, Introducing Hash Functions and Digital Signatures*

2
Introduction to Symmetric Encryption

After having an overview of cryptography, it's time now to present the principal algorithms in symmetric encryption and their logic and mathematical principles.

In *Chapter 1*, *Deep Diving into Cryptography*, we saw some symmetric cryptosystems such as **ROT13** and the **Vernam cipher**. Before going further into describing modern symmetric algorithms, we need an overview of the construction of block ciphers.

If you recall, symmetric encryption is performed through a key that is shared between the sender and receiver and vice versa. But how do we implement symmetric algorithms that are robust (in the sense of security) and easy to perform (computationally) at the same time? Let's see how we can answer this question by comparing asymmetric with symmetric encryption.

One of the main problems with asymmetric encryption is that it is not easy to perform the operations (especially the decryption), due to the high capacity of computation required to perform such algorithms at the recommended security levels. This problem implies that asymmetric encryption is not suitable for transmitting long messages, but it's better to exchange the key. Hence, by using symmetric encryption/decryption performed with the same shared key, we obtain a smoother scheme to exchange encrypted messages.

In this chapter, we will learn about the following topics:

- Understanding the basics of Boolean logic
- Understanding the basics of simplified DES
- Understanding and analyzing DES, Triple DES, and DESX
- Understanding AES (Rijndael) – the actual standard in symmetric encryption
- Implementing some logical and practical attacks on symmetric algorithms

By the end of the chapter, you will have understood how to implement, manage, and attack symmetric algorithms.

Notations and operations in Boolean logic

In order to understand the mechanism of symmetric algorithms, it is necessary to go over some notations in Boolean logic and these operations on a binary system.

As we have already seen in *Chapter 1*, *Deep Diving into Cryptography*, the binary system works with a set of bits of $\{0,1\}$. So, dealing with Boolean functions means performing logic calculations on a sequence of bits to generate an answer that could be either TRUE or FALSE.

The most frequently used functions are XOR (exclusive OR), OR (disjunction), and AND (conjunction). But there are a few other notations as well that will be explained soon.

A Boolean circuit aims to determine whether a variable, (x), combined with another variable, (y), satisfies the condition TRUE or FALSE. This problem is called the **Boolean Satisfiability Problem** (**SAT**) and it is of particular importance in computer science. SAT was the first problem to be shown as NP-complete. The question is as follows: given a certain function, does an assignment of the values TRUE or FALSE exist such that the expression results in TRUE?

A formula of *propositional logic* is *satisfiable* if there exists an assignment that can determine that a proposition is TRUE. If the result is FALSE for all possible variable assignments, then the proposition is said to be unsatisfiable. That is of great importance in algorithm theory, such as for the implementation of search engines, and even in hardware design or electronic circuits.

Let's give an example of propositional logic:

- **Premise 1**: *If the sky is clear, then it is sunny.*
- **Premise 2**: *There are no clouds in the sky.*
- **Conclusion**: *It's TRUE that it is sunny.*

As you can see in *Figure 2.1*, starting from an input and elaborating on the logic circuit with an algorithm, we obtain a conclusion of TRUE or FALSE.

All these concepts will be particularly useful in further chapters of the book, especially *Chapter 5, Introduction to Zero-Knowledge Protocols*, when we talk about **zero knowledge**, and *Chapter 9, Crypto Search Engine*, where we talk about a search engine that works with encrypted data:

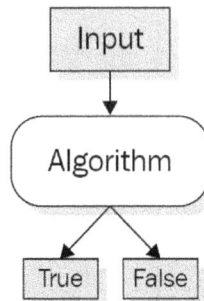

Figure 2.1 – A Boolean circuit gives two opposite variables as output

The basic operations performed in Boolean circuits are as follows:

- AND (**conjunction**): Denoted with the symbol $(X \wedge Y)$. This condition is satisfied when X together with Y is true. So, we are dealing with propositions such as pear AND apple, for example. If we are searching some content (let's say a database containing sentences and words), setting the AND operator will select all the elements containing both the words (pear *and* apple), not just one of them. Now let's explore how this operator works in mathematical mode. The AND operator transposed in mathematics is a multiplication of $(X * Y)$. The following is a representation of the *truth table* for all the logic combinations of the two elements. As you can see, only when $X * Y = 1$ does it mean that the condition of conjunction $(X \wedge Y)$ is satisfied:

Table of the TRUE for AND

x	y	x·y
0	0	0
0	1	0
1	0	0
1	1	1

Figure 2.2 – Mathematical table for "AND"

- OR (**disjunction**): Denoted by the symbol $(X \vee Y)$. This condition is satisfied when at least one of the elements of X or Y is true. So, we are dealing with a proposition such as pear OR apple. Our example of searching in a database will select all the elements containing at least one of the two words (pear *or* apple).

In the following table, you can see the OR operator transposed in the mathematical operation (X+Y). At least one of the variables assumes the value 1, so it satisfies the condition of disjunction (XVY), represented by the sum of the two variables:

Table of the TRUE for "OR"

x	y	x + y
0	0	0
0	1	1
1	0	1
1	1	1

Figure 2.3 – Mathematical table for "OR"

Important Note

Idempotence, from *idem + potence* (*same + power*), is a property of certain operations in mathematics and computer science that denotes that they can be applied multiple times without changing the result beyond the initial application. Boolean logic has idempotence within both AND and OR gates. A logical AND gate with two inputs of A will also have an output of A (1 AND 1 = 1, 0 AND 0 = 0). An OR gate has idempotence because 0 OR 0 = 0 and 1 OR 1 = 1.

- NOT (**negation**): Denoted with the symbol (¬X), meaning X excludes Y. So, we are dealing with propositions such as `pear NOT apple`. For example, if we search in a database, we are looking for documents containing only the first word or value (`pear`) and not for the second (`apple`). Finally, in the following table, you can see represented the NOT operator denoted by the symbol of negation, (¬X). It is represented by a unitary operation that gets back the opposite value with respect to its input:

Table of the TRUE for "NOT"

x	(not)x
0	1
1	0

Figure 2.4 – Mathematical table for "NOT"

These basic Boolean operators, AND, OR, and NOT, can be represented by a Venn diagram as follows:

Boolean AND, OR, and NOT

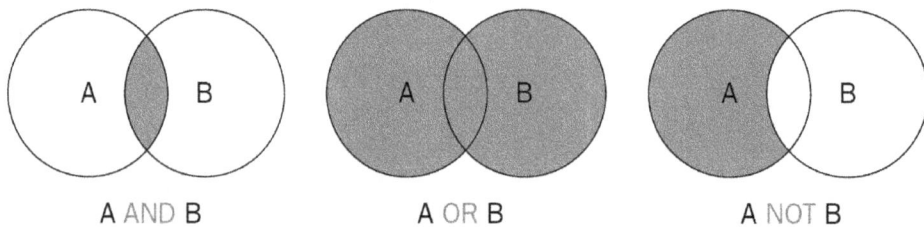

A AND B   A OR B   A NOT B

Figure 2.5 – Boolean operators represented by a Venn diagram

Besides the three basic operations just explored, there are more logic operations, including NAND, NOR, and XOR. All these operations are fundamental in cryptography. The NAND logical operator, for example, is used in **homomorphic encryption**; however, for now, we will limit ourselves to analyzing the XOR operator, also known as exclusive OR.

XOR is also denoted by the symbol $\oplus$.

The operation of A $\oplus$ B gives back the logic value of 1 if the number of variables that assume value 1 is odd. In other words, if we consider two variables, A and B, if both are either TRUE or FALSE, then the result is FALSE. As we can see in the following table, when A = 1 and B = 1, the result is 0 (FALSE).

In mathematical terms, XOR is an addition modulo 2, which means adding combinations of 1 and 0 in mod 2, as you can see in the following table, is called exclusive OR (often abbreviated to XOR):

XOR Table

A	$\oplus$	B	[A (XOR) B]
0	$\oplus$	0	= 0
0	$\oplus$	1	= 1
1	$\oplus$	0	= 1
1	$\oplus$	1	= 0

Figure 2.6 – Representing the XOR operations between 0 and 1

The XOR logic operator is used not only for cryptographic algorithms but also as a parity checker. If we run XOR in a logic circuit to check the parity bits in a word of 8 bits, it can verify whether the total number of ones in the word is a pair or not a pair.

Now that we have explored the operations behind Boolean logic, it's time to analyze the first algorithm of the symmetric family: DES.

DES algorithms

The first algorithm presented in this chapter is **Data Encryption Standard (DES)**. Its history began in 1973 when the **National Bureau of Standards (NBS)**, which later became the **National Institute of Standards and Technology (NIST)**, required an algorithm to adopt as a national standard. In 1974, IBM proposed **Lucifer**, a symmetric algorithm that was forwarded from NIST to the **National Security Agency (NSA)**. After analysis and some modifications, it was renamed DES. In 1977, DES was adopted as a national standard and it was largely used in electronic commerce environments, such as in the financial field, for data encryption.

Remarkable debates arose over the robustness of DES within the academic and professional community of cryptologists. The criticism derived from the short key length and the perplexity that, after a review advanced by the NSA, the algorithm could be subjected to a trapdoor, expressly injected by the NSA into DES to spy on encrypted communications.

Despite the criticisms, DES was approved as a federal standard in November 1976 and was published on January 15, 1977 as **FIPS PUB 46**, authorized for use on all unclassified data. It was subsequently reaffirmed as the standard in 1983, 1988 (revised as **FIPS-46-1**), 1993 (**FIPS-46-2**), and again in 1999 (**FIPS-46-3**), the latter prescribing **Triple DES** (also known as **3DES**, covered later in the chapter). On May 26, 2002, DES was finally superseded by the **Advanced Encryption Standard (AES)**, which I will explain later in this chapter, following a public competition. DES is a **block cipher**; this means that plaintext is first divided into blocks of 64 bits and each block is encrypted separately. The encryption process is also called the **Feistel system**, to honor *Horst Feistel*, one of the members of the team at IBM who developed Lucifer.

Now that a little bit of the history of this *progenitor* of modern symmetric algorithms has been revealed, we can go further into the explanation of its logical and mathematical scheme.

Simple DES

Simple DES is nothing but a simplified version of DES. Before we delve into how DES works, let's take a look into this simplest version of DES.

Just like DES, this simplified algorithm is also a block cipher, which means that plaintext is first divided into blocks. Because each block is encrypted separately, we are supposed to analyze only one block.

The key, [K], is made up of 9 bits and the message, [M], is made up of 12 bits.

The main part of the algorithm, just like in DES, is the **S-Box**, where **S** stands for **Substitution**. Here lies the true complexity and non-linear function of symmetric algorithms. The rest of the algorithm is only permutations and shifts over the bits, something that a normal computer can do automatically, so there is no reason to go crazy over it.

An S-Box in this case is a 4X16 matrix consisting of 6 bits as input and 4 bits as output.

We will find that the S-Box is present in all modern symmetric encryption algorithms, such as DES, Triple DES, Bluefish, and AES.

The four rows are represented by progressive 2 bits, as follows:

```
00
01
10
11
```

The 16 lines of the columns instead consist of 4 bits in this sequence:

```
0000 0001 0010 ...... ...... ...... ...... 1111
```

The matrix's boxes consist of random numbers between 0 and 15, which means they never get repeated inside the same row.

In order to better understand how S-Box is implemented and how it works, here is an example: 011011. This string of bits has two outer bits, 01, and four middle bits, 1101.

In this case (working in the binary system, using the *N*2 notation), $(01)_2$ corresponds to the second row, and $(1101)_2$ corresponds to the 13th column. By finding the intersection of the column and the row, we obtain $(1001)_2$.

You can see the representation in binary numbers of the S-Box matrix described here:

```
            0    1    2...                          13    14    15

          0000 0001 0010 .....       .........  ............1101 1110 1111
                                                               |
   1 00                                                        ↓
   2 01  - - - - - - - - - - - - - - - - - - - - - - - - - →  1001
   3 10
   4 11
```

Figure 2.7 – An S-Box matrix (intersection) of 4X16 represented in binary numbers

The same matrix can be represented in decimal numbers as follows:

0	1	2	3	4	5	6	7	8	9	10	11	12	13	14	15
2	12	4	1	7	10	11	6	8	5	3	15	13	0	14	9
14	11	2	12	4	7	13	1	5	0	15	10	3	9	8	6
4	2	1	11	10	13	7	8	15	9	12	5	6	3	0	14
11	8	12	7	11	14	2	3	6	15	0	9	10	4	5	3

Figure 2.8 – An S-Box matrix of 4X16 represented in decimal numbers

Here, the number 9 represents the intersection between row 2 and column 13. So, the number found crossing row 2 and column 13, represented in a binary system as $(1001)_2$, corresponds to 9 in the decimal system.

Now that we are clear on what S-Box is and how it is designed, we can see how the algorithm works.

Bit initialization

The message, M, consisting of 12 bits, is divided into two parts, L_0 and R_0, where L_0, the left half, consists of the first 6 bits and R_0, the right half, consists of the last 6 bits:

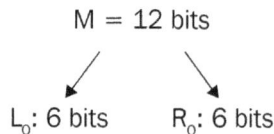

$$M = 12 \text{ bits}$$

$$L_0: 6 \text{ bits} \qquad R_0: 6 \text{ bits}$$

Figure 2.9 – Message (M) is split into 6 bits to the left and 6 bits to the right

Now that we have a clear concept of S-Box and bit initialization, let's proceed with the other phases of the process: bit expansion, key generation, and bit encryption.

Bit expansion

Each block of bits, the left and right parts, is expanded through a particular function that is normally called *f*.

The DES algorithm uses an expansion at 8 bits (1 byte) starting from 6-bit input for each block of plaintext.

Moreover, DES uses a modality of partition called **Electronic Codebook (ECB)** to divide the 64 bits of plaintext into 8X8 bits, for each block performing the (E_k) encryption function.

Any *f* could be differently implemented, but just to give you an example, the first input bit becomes the first output, the third bit becomes the fourth and the sixth, and so on. Just like the following example, let's say we want to expand the 6-bit L_0: 011001 input with an expansion function, Exp, following this pattern:

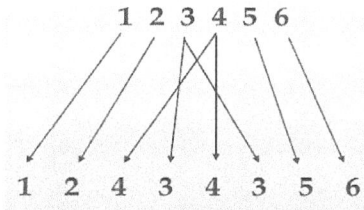

Figure 2.10 – Bit expansion function [EXP]

As you can see in the preceding figure, L_0 = (011001)$_2$ has been expanded with f [12434356].

Then, L_0: 011001 will be expanded into (01010101)$_2$, as shown in the following figure:

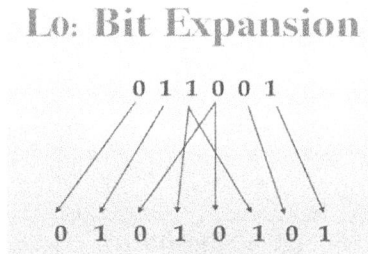

Figure 2.11 – L_0 (011001)$_2$ bit expansion 8 bits

Expanding the 6-bit R_0: $(100110)_2$ input to 8 bits with the same pattern, f [12434356], in R_{i-1} = $(100110)_2$, we obtain the following:

Ro: Bit Expansion

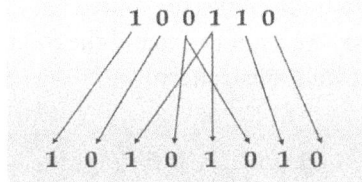

1 0 0 1 1 0

1 0 1 0 1 0 1 0

Figure 2.12 – R_0 $(100110)_2$ bit expansion 8 bits

So, the expansion of R_{i-1} will be $(10101010)_2$.

Key generation

As we have already said, the master key, [K], is made up of 9 bits. For each round, we have a different encryption key, [K_i], generated by 8 bits of the master key, starting counting from the ith round of encryption.

Let's take an example to clarify the key generation K_4 (related to the fourth round):

- K = 010011001 (9-bit key, the master key)

- K_4 = 01100101 (8 bits taken by K)

The following figure will help you better understand the process:

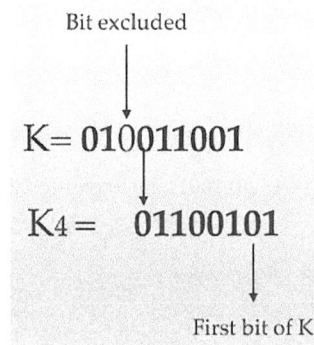

Bit excluded

K= 010011001

K4 =   01100101

First bit of K

Figure 2.13 – Example of key generation

As you can see in the previous figure, we are processing the fourth round of encryption, so we start to count from the fourth bit of master key [K] to generate [K4].

Bit encryption

To perform the bit encryption, (E), we use the XOR function between $R_{i-1} = (100110)2$ expanded and $K_i = (01100101)_2$.

I call this output $E(K_i)$:

```
Exp(R_i-1) ⊕ Ki = 10101010 ⊕ 01100101 = E(K_i) (11001111)
```

At this point, we split $E(K_i)$, consisting of 8 bits, into two parts: a 4-bit half for the left and a 4-bit half for the right:

```
L(EKi) = (1100)_2                    R(EK_i) = (1111)_2.
```

Now, we process the 4 bits to the left and the 4 bits to the right with two S-Box 2X8 matrices consisting of 3 bits for each element. The input, as I mentioned earlier, is 4 bits: the first one is the row and the last three represent a binary number to indicate the column (the same as previously, just with fewer bits). So, 0 stands for first and 1 stands for second. Similarly, 000 stands for the first column, 001 stands for the second column, and so on until 111.

We call the two S-Boxes S1 and S2. The following figure represents the elements of each one:

	101	010	001	110	011	100	111	000
S1	001	100	110	010	**000**	111	101	011
S2	100	000	110	101	111	001	011	010
	101	011	000	111	110	010	001	**100**

Figure 2.14 – Example of S-Boxes

$L(E(K_i)) = (1100)_2$ is processed by S1; so, the element of the second row, $(1)_2$, and the fourth column, $(100)_2$, is the output, here represented by the number $(000)_2$.

$R(E(K_i)) = (1111)_2$ is processed by S2; so, the element of the second row, $(1)_2$, and the seventh column, $(111)_2$, is the output, here represented by the number $(100)_2$.

Now, the last step is the concatenation of the two outputs obtained, here expressed by the notation ||, which will perform the ciphertext:

```
S1(L(E(K_i))) = (000)_2  ||   S2(R(E(K_i))) = (100)_2
000 || 100 = (000100)_2 f (R_i-1, K_i) = (000100)_2
```

The following figure shows how the encryption of the first round (the right side) of the function f mathematically works:

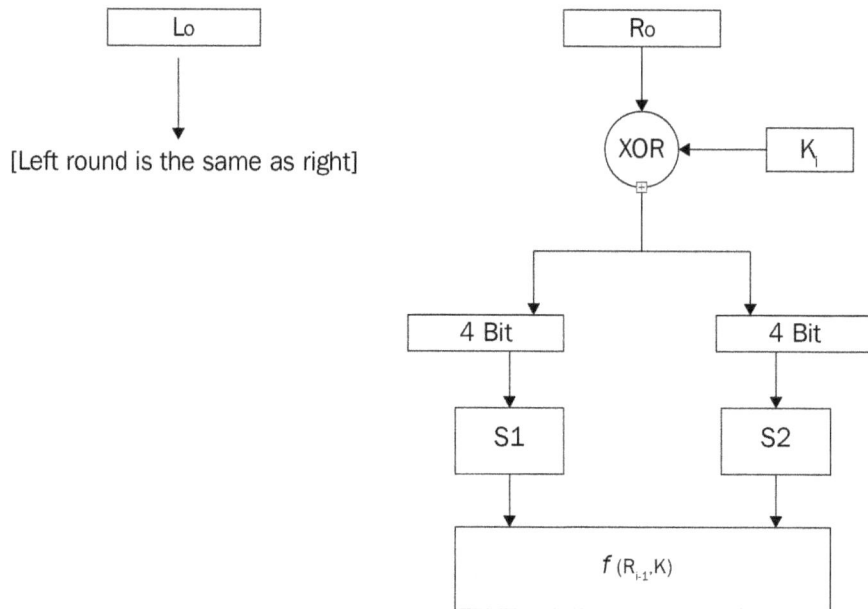

Figure 2.15 – Mathematical scheme of (simple) DES encryption at the first round (right side)

Now that we have understood how **simple DES** works and covered the basics of symmetric encryption, it will be easier to understand how the DES family of algorithms work.

As you have seen, the combination of permutations, XOR and **shift**, is the pillar of the structure of the **Feistel system**.

DES

DES is a 16-round encryption/decryption symmetric algorithm. DES is a 64-bit cipher in every sense. The operations are performed by dividing the message, [M], into 64-bit blocks. The key is also 64 bits; however, it is effectively 56 bits (plus 8 bits for parity: 8th, 16th, 24th...). This technique eventually allows us to check errors. Finally, the output, (c), is 64 bits too.

I would like you to focus on the DES encryption scheme of *Figure 2.15* to fully understand DES encryption.

Key generation in DES

Based on *Shannon's principle of confusion and diffusion*, DES, just like most symmetric algorithms, operates over bit scrambling to obtain these two effects.

As already mentioned, the DES master key is a 64-bit key. The key's bits are enumerated from 1 to 64, where every eighth bit is ignored, as you can see in the highlighted column in the following table:

Input Key in DES

```
 1  2  3  4  5  6  7  8
 9 10 11 12 13 14 15 16
17 18 19 20 21 22 23 24
25 26 27 28 29 30 31 32
33 34 35 36 37 38 39 40
41 42 43 44 45 46 47 48
49 50 51 52 53 54 55 56
57 58 59 60 61 62 63 64
```

Figure 2.16 – Bits deselected in the DES master key

After the deselection of the bits, the new key is a 56-bit key.

At this point, the first permutation on the 56-bit key is computed. The result of this operation is *confusion* on the bit positions; then, the key is divided into two 28-bit sub-keys called $C0$ and $D0$.

After this operation (always in the same line to create a bit of confusion and diffusion), it performs a circular shift process as shown in the following table:

Round	1	2	3	4	5	6	7	8	9	10	11	12	13	14	15	16
Number of bits shifted	1	1	2	2	2	2	2	2	1	2	2	2	2	2	2	1

Figure 2.17 – Showing the number of key bits shifted in each round in DES

If you look at the preceding table, you can notice rounds 1, 2, 9, and 16 shift left by only 1 bit; all the other rounds shift left by 2 bits.

Let's take as an example of C0, D0 (the original division of the key in 28-bit left and 28-bit right), expressed in binary notation as follows:

```
C0 = 1111000011001100101010101001
D0 = 0101010101100110011110001111
```

Now, from C0 and D0, C1 and D1 will be generated, as follows:

```
C1 = 1110000110011001010101010011
D1 = 1010101011001100111100011110
```

If you focus on the step of the generation of C0 --> C1, you can better understand how it works – it's a simple shift to the left of all the bits of C0 with respect to C1:

```
C0 = 1111000011001100101010101001
C1 = 1110000110011001010101010011
```

After the circular shift, as explained, the next step is to process a selection of 48 bits over the subset key of 56 bits. It's a simple permutation of position: just to give an example, bit number 14 moves to the first position and bit number 32 moves to the last position (48th). As you can see in the following table, some bits, just like bit number 18, are discarded in the new configuration, so you don't find them in the table. At the end of the process of bit compression, only 48 bits are selected; consequently, 8 bits are discarded:

14	17	11	24	1	5	3	28	15	6	21	10
23	19	12	4	26	8	16	7	27	20	13	2
41	52	31	37	47	55	30	40	51	45	33	48
44	49	39	56	34	53	46	42	50	36	29	32

Figure 2.18 – Transformation and compression in a 48-bit subset key

In the following figure, you can see the whole process of **key generation**, which combines **Parity Drop**, **Shift Left**, and **Compression**:

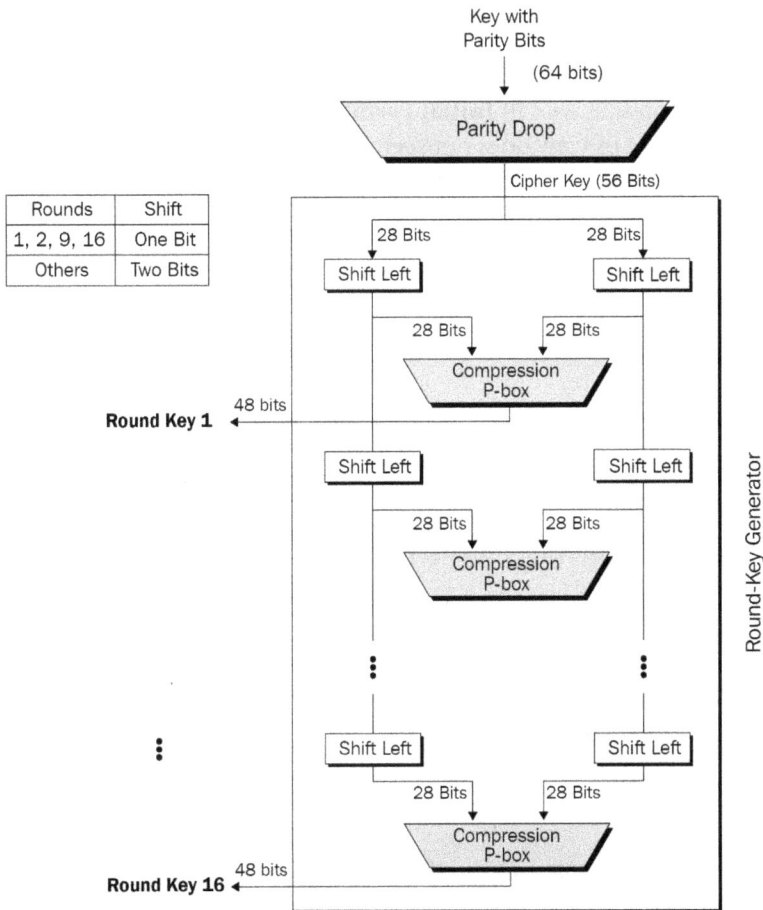

Rounds	Shift
1, 2, 9, 16	One Bit
Others	Two Bits

Figure 2.19 – The key generation scheme

Because of this compression/confusion/permutation technique, DES is able to determine different sub-keys, one per round of 48 bits. This makes DES difficult to crack.

Encryption

After we have generated the key, we can proceed with the encryption of the message, [M].

The encryption scheme of DES consists of three phases:

- **Step 1 – initial permutation**: First, the bits of the message, [M], are permutated through a function that we call **Initial Permutation (IP)**. This operation from a cryptographic point of view does not seem to augment the security of the algorithm. After the permutation, the 64 bits are divided into 32 bits in L0 and 32 bits in R0 just like we did in simplified DES.

- **Step 2 – rounds of encryption**: For $0 \leq i \leq 16$, the following operations are executed:

  ```
  Li = Ri-1
  ```

  Compute Ri = Li-1 $\oplus$ f(Ri-1, Ki).

  Here, Ki is a string of 48 bits obtained from the key K (round key *j*) and f is a function of expansion similar to the f described earlier for simple DES.

  Basically, for $i = 1$ (the first round), we have the following:

  ```
  L1 = R0
  R1 = L0 ⊕ f(R0, K1)
  ```

- **Step 3 – final permutation**: The last part of the algorithm at the 16th round (the last one) consists of the following:

  a. Exchanging the left part, L16, with the right part, R16, in order to obtain (R16, L16)

  b. Applying the inverse, INV, of the IP to obtain the ciphertext, c, where c = INV(IP(R16, L16))

The following figure is a representation of an intelligible scheme of DES encryption:

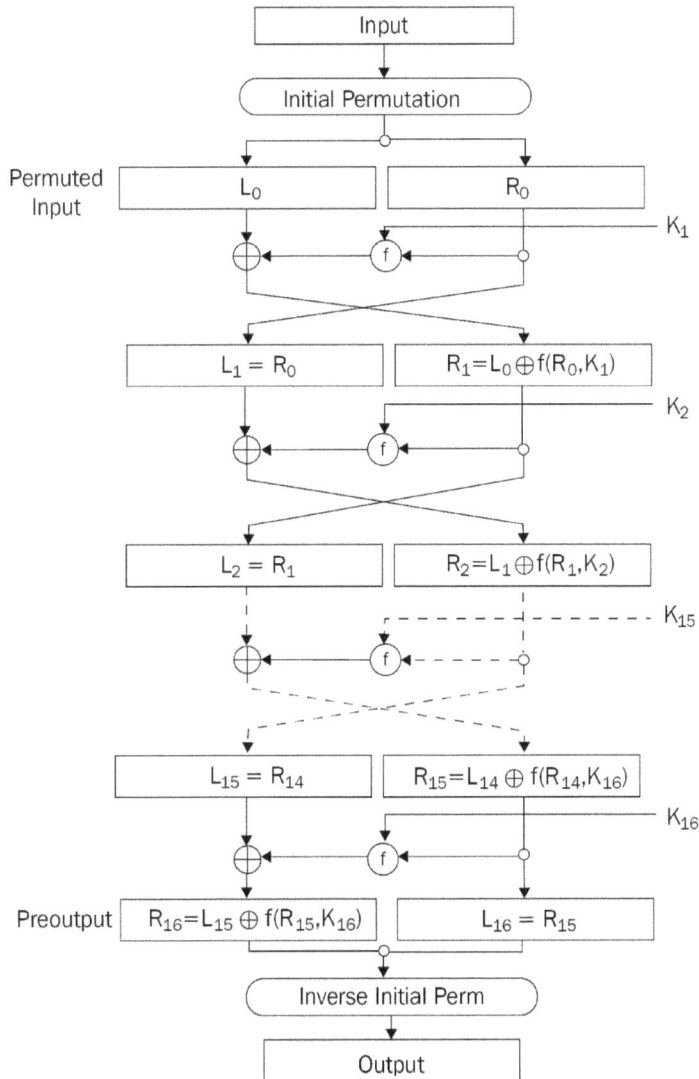

Figure 2.20 – DES encryption

To summarize the encryption stage in DES, we performed a complex process for key generation, where a selection of 48-bit subsection keys on a master key of 64 bits was made. There are consequently three steps: IP, rounds of encryption, and final permutation.

Now that we have analyzed the encryption process, we can move on to DES decryption analysis.

Decryption

DES decryption is very easy to understand. Indeed, to get back the plaintext, we perform an inverse process of encryption.

The decryption is performed in exactly the same manner as encryption, but by inverting the order of the keys ($K_1...K_{16}$) so that it becomes ($K_{16}...K_1$).

In the following figure, you see the decryption process in a flow chart scheme:

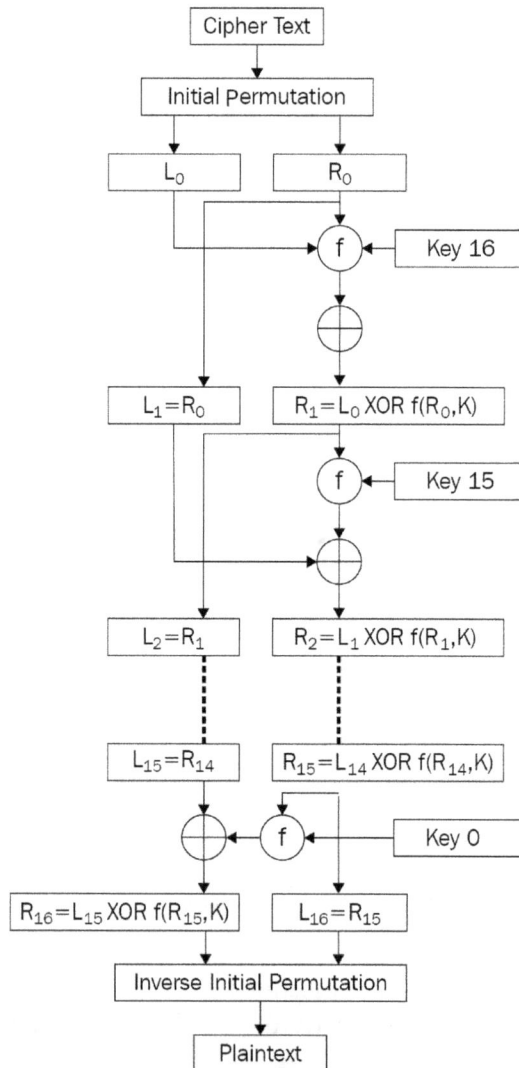

Figure 2.21 – DES decryption process

So, to describe the decryption process: you take the ciphertext and operate the first IP on it, then `XOR L0` (left part) with `R0 = f(R0, K16)`.

Then you keep on going like that for each round, make a final permutation, and end up finding the plaintext.

Now that we have arrived at the end of the DES algorithm process, let's go ahead with the analysis of the algorithm and its vulnerabilities.

Analysis of the DES algorithm

Going a little bit more into the details of the algorithm, we can discover some interesting things.

One of the most interesting steps of DES is the `XOR` operation performed between the sub-keys (`K1, K2...K16`) and the half part of the message, `[M]`, at each round.

In this step, we find the S-Box: the function f previously described in simplified DES.

As we saw earlier, the S-Box is a particular matrix in DES that consists of 4 rows and 16 columns (S-Box 4X16) fixed by the NSA:

i									S_1							
1	14	4	13	1	2	15	11	8	3	10	6	12	5	9	0	7
	0	15	7	4	14	2	13	1	10	6	12	11	9	5	3	8
	4	1	14	8	13	6	2	11	15	12	9	7	3	10	5	0
	15	12	8	2	4	9	1	7	5	11	3	14	10	0	6	13
2	15	1	8	14	6	11	3	4	9	7	2	13	12	0	5	10
	3	13	4	7	15	2	8	14	12	0	1	10	6	9	11	5
	0	14	7	11	10	4	13	1	5	8	12	6	9	3	2	15
	13	8	10	1	3	15	4	2	11	6	7	12	0	5	14	9
3	10	0	9	14	6	3	15	5	1	13	12	7	11	4	2	8
	13	7	0	9	3	4	6	10	2	8	5	14	12	11	15	1
	13	6	4	9	8	15	3	0	11	1	2	12	5	10	14	7
	1	10	13	0	6	9	8	7	4	15	14	3	11	5	2	12
4	7	13	14	3	0	6	9	10	1	2	8	5	11	12	4	15
	13	8	11	5	6	15	0	3	4	7	2	12	1	10	14	9
	10	6	9	0	12	11	7	13	15	1	3	14	5	2	8	4
	3	15	0	6	10	1	13	8	9	4	5	11	12	7	2	14
5	2	12	4	1	7	10	11	6	8	5	3	15	13	0	14	9
	14	11	2	12	4	7	13	1	5	0	15	10	3	9	8	6
	4	2	1	11	10	13	7	8	15	9	12	5	6	3	0	14
	11	8	12	7	1	14	2	13	6	15	0	9	10	4	5	3
6	12	1	10	15	9	2	6	8	0	13	3	4	14	7	5	11
	10	15	4	2	7	12	9	5	6	1	13	14	0	11	3	8
	9	14	15	5	2	8	12	3	7	0	4	10	1	13	11	6
	4	3	2	12	9	5	15	10	11	14	1	7	6	0	8	13
7	4	11	2	14	15	0	8	13	3	12	9	7	5	10	6	1
	13	0	11	7	4	9	1	10	14	3	5	12	2	15	8	6
	1	4	11	13	12	3	7	14	10	15	6	8	0	5	9	2
	6	11	13	8	1	4	10	7	9	5	0	15	14	2	3	12
8	13	2	8	4	6	15	11	1	10	9	3	14	5	0	12	7
	1	15	13	8	10	3	7	4	12	5	6	11	0	14	9	2
	7	11	4	1	9	12	14	2	0	6	10	13	15	3	5	8
	2	1	14	7	4	10	8	13	15	12	9	0	3	5	6	11

Figure 2.22 – S-Box matrix in DES

Take a look at the specifics of the S-Box in the fifth round of DES:

```
2, 12, 4, 1, 7, 10, 11, 6, 8, 5, 3, 15, 13, 0 14, 9
14, 11, 2, 12, 4, 7, 13, 1, 5, 0, 15, 10, 3, 9, 8, 6
4, 2, 1, 11, 10, 13, 7, 8, 15, 9, 12, 5, 6, 3, 0, 14
11, 8, 12, 7, 11, 14, 2, 13, 6, 15, 0, 9, 10, 4, 5, 3
```

As you might notice, the numbers included are between 0 and (R_{i-1}) $16-1 = 15$.

So, 48 bits of input will give exactly 48 bits of output after the XOR operation is performed with $[K_i]$.

Moreover, if you observe carefully, in the 14th column, all the numbers are very low: 0, 9, 3, and 4. *Would this combination pose a problem for security?*

You will be perplexed if I tell you that it will not be an issue to play with little numbers inside an S-Box.

Another question that may come to you spontaneously might be *Why is the key only 56 bits and not 64 bits?* Because, as I already mentioned, the other 8 bits are used for pairing.

Actually, the initial master key is 64 bits in length, so every eighth bit of the key is discarded. The final result is a 56-bit key, as you see in the following figure:

1	2	3	4	5	6	7	8	9	10	11	12	13	14	15	16
17	18	19	20	21	22	23	24	25	26	27	28	29	30	31	31
33	34	35	36	37	38	39	40	41	42	43	44	45	46	47	48
49	50	51	52	53	54	55	56	57	58	59	60	61	62	63	64

Figure 2.23 – Bit discarded in the DES key generation algorithm

There is one more concern that could arise from the method of encryption adopted in DES. After the IP, as you can see, the bits are encrypted only on the right side through the $f(R_{i-1}, K_i)$ function. You might ask whether this is less secure than encrypting all the bits. If you analyze the scheme properly, you will notice that at each round, the bits are exchanged from left to right, then encrypted, and vice versa. This technique is like a wrap that allows all the bits to be encrypted, not just the right part as it would seem at first glance.

Looking back at the initial and final permutation functions, you may ask: when making an initial and final inverse permutation, isn't the final result neutral? As I already mentioned, there isn't any cryptographic sense in performing a permutation of bits like that. The reason is that bit insertion into hardware in the 70s was much more complicated than it is now. To complete the discussion, I can say that the entire process adopted in DES for substitution, permutation, E-expansion, and bit-shifting generates *confusion* and *diffusion*. At the beginning of this section, I already quoted this concept when I mentioned the security cipher principle identified by *Claude Shannon* in his *A Mathematical Theory of Communication*.

Violation of DES

The history of the attacks performed to crack DES since its creation is rich in anecdotes. In 1975, among the academic community, skepticism against the robustness of the key length with respect to the 56-bit keys started to arise. Many articles have been published; one very interesting prediction of *Whitney Diffie* and *Martin Hellman* (the same pair from the *Diffie and Hellman exchange of the key* seen in *Chapter 3, Asymmetric Encryption*) was that a computer worth $20 million (in 1977) could be built to break DES in only 1 day.

More than 20 years later, in 1998, the **Electronic Frontier Foundation** (**EFF**) developed a dedicated computer called **DES Cracker** to break DES. The EFF spent a little less than $250,000 and employed 37,050 units embedded into 26 electronic boards. After 56 hours, the supercomputer gained the decryption of the plaintext message object of a challenge as there was a $10,000 reward for the decryption of the ciphertext. The method adopted was a simple *brute-force* method to analyze all the possible combinations of bits given by the 2^{56} (about 72 quadrillion) possible keys. The EFF was able to crack DES using hardware that incorporated 1,500 microchips working at 40 MHz, in 4.5 days of running time. Imagine, only 1 microchip would take 38 years to explore the entire set of keys.

At this point, the authorities decided to replace the algorithm with a new symmetric key algorithm, and here came AES. But before exploring AES, let's analyze some possible attacks on DES.

I will present next some possible attacks on DES, taking into consideration that most of these methods are used to attack most symmetric algorithms. Some of the attacks are specific to blocking ciphers, while others are valid for streaming ciphers too. The difference is that in a stream cipher, 1 byte is encrypted at a time, while in a block cipher, ~128 bits are encrypted at a time (block):

- **Brute-force attack**: This basic method of attack can be performed for any known cipher, meaning trying all the possibilities to find the key. If you recall, the key length of DES is a 56-bit key, which means an attacker would evaluate all the 72,057,594,037,927,936 possibilities ($2^{\wedge}56$). This is not a computation to be taken lightly, but despite this, DES has been a breakable algorithm since the early 90s.

- **Linear cryptanalysis**: This is essentially a statistical method of attack based on known plaintext. It doesn't guarantee success every time, but it does work most of the time. The idea is to start from the known input (plaintext) and arrive at determining the key of encryption and, consequently, all the other outputs generated by that key.

- **Differential cryptanalysis**: This method is technical and requires observing some vulnerabilities inside DES (similar to other symmetric algorithms). This attack method attempts to discover the plaintext or the key, starting from a chosen plaintext. Unlike linear cryptanalysis, which starts from improbable known plaintext, the attacker operates knowing chosen plaintext.

Last but not least, a vulnerability of DES is called *weak keys*: these keys are simply not able to perform any encryption. This is very dangerous because if applied, you get back plaintext. These keys are well known in cryptography and have to be avoided.

That happens when the sequence of the 16th key (during the key generation) produces all 16 identical keys.

Let's see an example of this problem:

- **A sequence of bits all equal to** 0000000000000000 or 1111111111111111
- **A sequence of alternate bits**, 010101010101 or 1010101010101010

In all four cases, it turns out that the encryption is auto-reversible, or in other words, if you perform two encryptions on the same ciphertext, you will obtain the original plaintext.

Triple DES

As I mentioned previously, one of the main weaknesses found in DES was the key length of 56 bits. So, to amplify the volume of keys and to extend their life, a new version of DES was proposed in the form of **Triple DES**.

The logic behind 3DES is the same as DES; the difference is that here we run the algorithm three times with three different keys.

The following figure shows a scheme proposed to better understand 3DES:

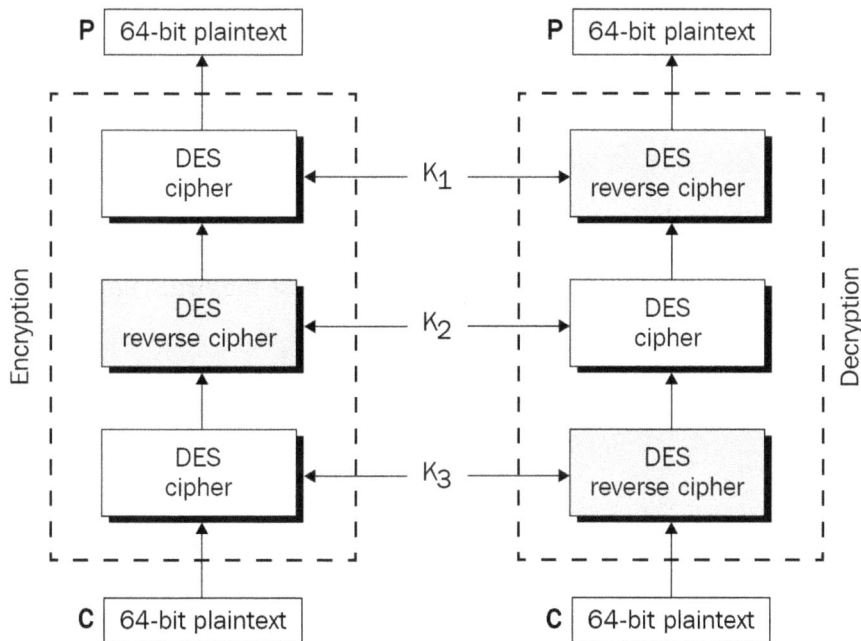

Figure 2.24 – Triple DES encryption/decryption scheme

Let's see how the encryption and decryption stages work in DES, based on the scheme illustrated in the preceding figure.

Encryption in 3DES works as follows:

1. Encrypt the plaintext blocks using single DES with the $[K_1]$ key.
2. Now, decrypt the output of *step 1* using single DES with the $[K_2]$ key.
3. Finally, encrypt the output of *step 2* using single DES with the $[K_3]$ key.

The output of *step 3* is the ciphertext (C).

Decryption in 3DES

The decryption of ciphertext is a reverse process. The user first decrypts using [K$_3$], then decrypts with [K$_2$], and finally decrypts with [K$_1$].

DESX

The last algorithm of the DES family is **DESX**. This is a reinforcement of DES's key proposed by *Ronald Rivest* (the same coauthor of RSA).

Given that DES encryption/decryption remains the same as earlier, there are three chosen keys: [K$_1$], [K$_2$], and [K$_3$].

The following encryption is performed:

```
C = [K₃] ⊕ EK₁ ([K₂] ⊕ [M])
```

First, we have to perform the encryption (E$_K$), making an XOR between [K$_2$] and the message, [M]. Then, we apply DES, encrypting with [K$_1$] 56 bits. Finally, we add the outputs E$_{K1}$, XOR, and [K$_3$]. This method allows us to increase the virtual key to be 64 + 56 + 64 = 184 bits, instead of the normal 56 bits:

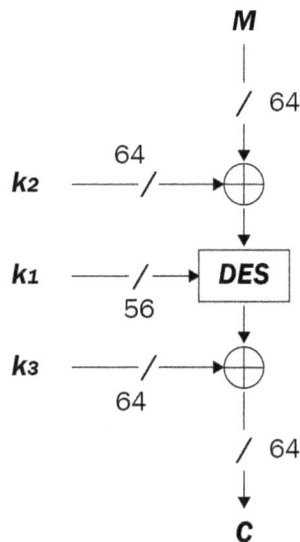

Figure 2.25 – DESX encryption scheme

After exploring the DES, 3DES, and DESX algorithms, we will approach another pillar of the symmetric encryption algorithm: AES.

AES Rijndael

AES, also known as **Rijndael**, was chosen as a very robust algorithm by NIST (the US government) in 2001 after a 3-year testing period among the cryptologist community.

Among the 15 candidates who competed for the best algorithm, there were five finalists chosen: **MARS (IBM)**, **RC6 (RSA Laboratories)**, **Rijndael (Joan Daemen and Vincent Rijmen)**, **Serpent (Ross Anderson and others)**, and **Twofish (Bruce Schneier and others)**. All the candidates were very strong but, in the end, Rijndael was a clear winner.

The first curious question is about its name: how is Rijndael pronounced?

It's dubiously difficult to pronounce this name. From the web page of the two authors, we can read that there are a few ways to pronounce this name depending on the nationality and the mother tongue of who pronounces it.

Just to start, I can say that AES is a block cipher, so it can be performed in different modes: ECB (already seen in DES), **Cipher Block Chaining (CBC)**, **Cypher Feedback Block (CFB)**, **Output Feedback Block (OFB)**, and **Counter (CTR)** mode. We will see some better differences between implementations in this section.

AES can be performed using different key sizes: 128-, 192-, and 256-bit. NIST's competition aimed to find an algorithm with some very strong characteristics, such as it should be operating in blocks of 128 bits of input or it should be able to be used on different kinds of hardware, from 8-bit processors (used also in a smart card) to 32-bit architectures, commonly adopted in personal computers. Finally, it should be fast and very robust.

Under certain conditions (that you will discover later), I think this is one of the best algorithms ever; indeed, I have chosen to implement AES 256 in our **Crypto Search Engine (CSE)**. We will see CSE again in *Chapter 9, Crypto Search Engine*. We adopt AES to secure the symmetric encryption of data encrypted and transmitted between virtual machines that encrypt and store data.

Description of AES

Discussing AES would alone require a dedicated chapter. In this section, I provide an overview of the algorithm. For those of you interested in knowing more, you can refer to the documentation presented by NIST (published on November 26, 2001) reported in the document titled *FIPS PUB 197*. I am limited in this chapter to describing the algorithm at just a high level and will give my comments and suggestions.

Most importantly, to avoid any confusion, I will analyze AES in a different manner not found in other papers. My analysis of AES will be based on the subdivision of the algorithm into different steps. I have called these steps **Key Expansion** and **First Add Round Key**; then, as you will see later on, each step is divided into four sub-steps, called **SubBytes transformation**, **ShiftRows transformation**, **MixColumn**, and **AddRoundKey**. The important thing is to understand the scheme of the algorithm, then each round works similarly, and you can easily be guided to understand the mechanism of 10 rounds for a 128-bit key, 12 rounds for 192 bits, and 14 rounds for 256 bits.

Key Expansion (KE) works as follows:

1. The fixed key input of 128 bits is expanded into a key length depending on the size of AES: 128, 192, or 256.

2. Then, the $[K_1]$, $[K_2]$,... $[K_r]$ sub-keys are created to encrypt each round (generally adding XOR to the round).

3. AES uses a particular method called Rijndael's *key schedule* to expand a short master key to a certain number of round keys.

First Add Round Key (F-ARK) works as follows.

It is the first operation. The algorithm takes the first key, $[K_1]$, and adds it to AddRoundKey: using a bitwise XOR of the current block with a portion of the expanded key.

Rounds R_1 to R_{n-1} work as follows.

Each round (except the last one) is divided into four steps called layers consisting of the following:

1. **SubBytes (SB) transformation**: This step is a fundamental non-linear step, executed through a particular S-Box (we have already seen how an S-Box works in DES). You can see the AES S-Box in the following figure.

2. **ShiftRows (SR) transformation**: This is a scrambling of a bit that causes diffusion on multiple rounds.

3. **MixColumns (MC) transformation**: This step has a similar scope to SR but is applied to the columns.

4. **AddRoundKey (ARK)**: The round key is XORed with the result of the previous layer.

The following figure represents S-Box Rijndael expressed in hexadecimal notation:

		0	1	2	3	4	5	6	7	8	9	a	b	c	d	e	f
	0	63	7C	77	7B	F2	6B	6F	C5	30	01	67	2B	FE	D7	AB	76
	1	CA	82	C9	7D	FA	59	47	F0	AD	D4	A2	AF	9C	A4	72	C0
	2	B7	FD	93	26	36	3F	F7	CC	34	A5	E5	F1	71	D8	31	15
	3	04	C7	23	C3	18	96	25	9A	07	12	80	E2	EB	27	B2	75
	4	09	83	2C	1A	1B	6E	5A	A0	52	3B	D6	B3	29	E3	2F	84
	5	53	D1	00	ED	20	FC	B1	5B	6A	CB	BE	39	4A	4C	58	CF
	6	D0	EF	AA	FB	43	4D	33	85	45	F9	02	7F	50	3C	9F	A8
X	7	51	A3	40	8F	92	9D	38	F5	BC	B6	DA	21	10	FF	F3	D2
	8	CD	0C	13	EC	5F	97	44	17	C4	A7	7E	3D	64	5D	19	73
	9	60	81	4F	DC	22	2A	90	88	46	EE	B8	14	DE	5E	0B	DB
	a	E0	32	3A	0A	49	06	24	5C	C2	D3	AC	62	91	95	E4	79
	b	E7	C8	37	6D	8D	D5	4E	A9	6C	56	F4	EA	65	7A	AE	08
	c	BA	78	25	2E	1C	A6	B4	C6	E8	DD	74	1F	4B	BD	8B	8A
	d	70	3E	B5	66	48	03	F6	0E	61	35	57	B9	86	C1	1D	9E
	e	E1	F8	98	11	69	D9	8E	94	9B	1E	87	E9	CE	55	28	DF
	f	8C	A1	89	0D	BF	E6	42	68	41	99	2D	0F	B0	54	BB	16

Figure 2.26 – S-Box Rijndael

The last round runs all the operations of the previous rounds except for layer 3: **MC**. After all the earlier-mentioned processes are run for each round n times (depending on the size of the key: 14 rounds if the key is 256 bits), AES encryption obtains the ciphertext, (C), as shown in the following figure:

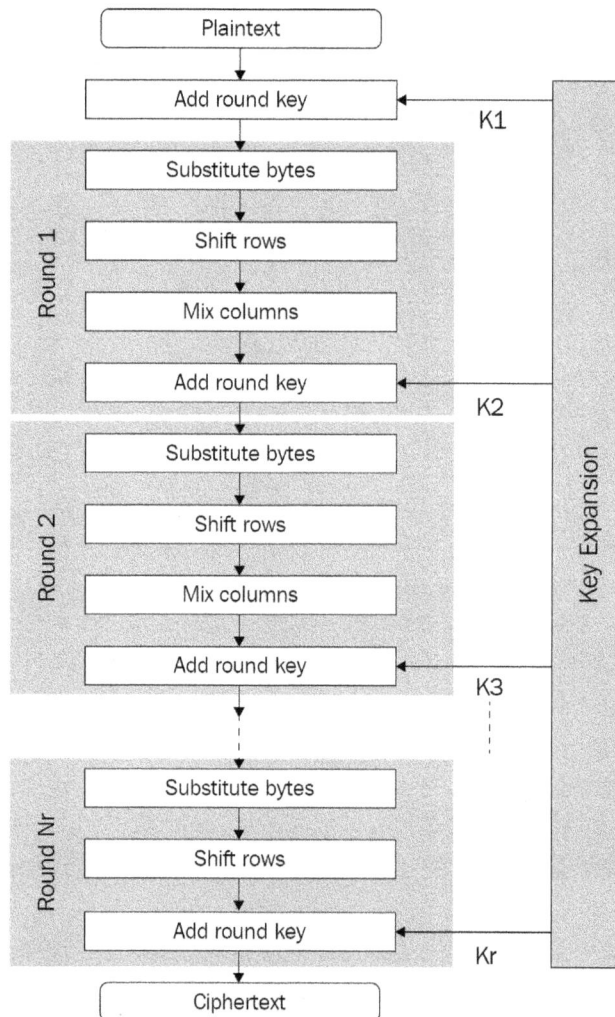

Figure 2.27 – Encryption scheme in AES

Thus, we can re-schematize the entire process of AES encryption in a mathematical function as follows:

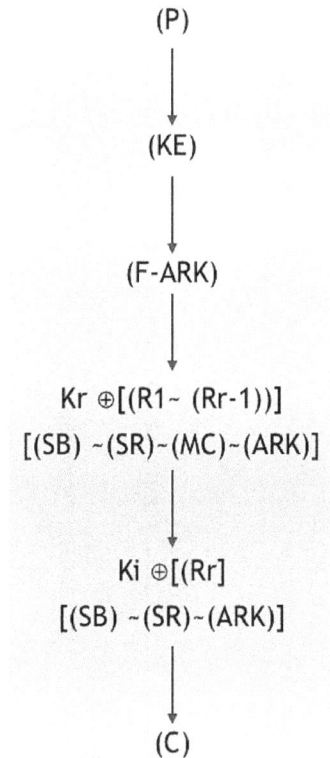

(P)

↓

(KE)

↓

(F-ARK)

↓

Kr ⊕[(R1~ (Rr-1))]

[(SB) ~(SR)~(MC)~(ARK)]

↓

Ki ⊕[(Rr]

[(SB) ~(SR)~(ARK)]

↓

(C)

Figure 2.28 – Re-schematizing an AES flow chart with mathematical functions

As you can notice from the proposed scheme in the preceding figure, we have the following:

- K_r ⊕ [(R_1~ (R_{r-1}))] represents all the mathematical processes performed between each round key, (K_r), XORed with *the inside functions* of each rounds, starting from the first round (after the F-ARK to the last round (excluded)). So, inside [(R1~ (Rr-1))], we find the [(SB) ~(SR) ~ (MC) ~ (ARK)] functions.

- In the last round, as you can notice, MC is not present.

In the following figure, you can see the entire process of encryption and decryption in AES:

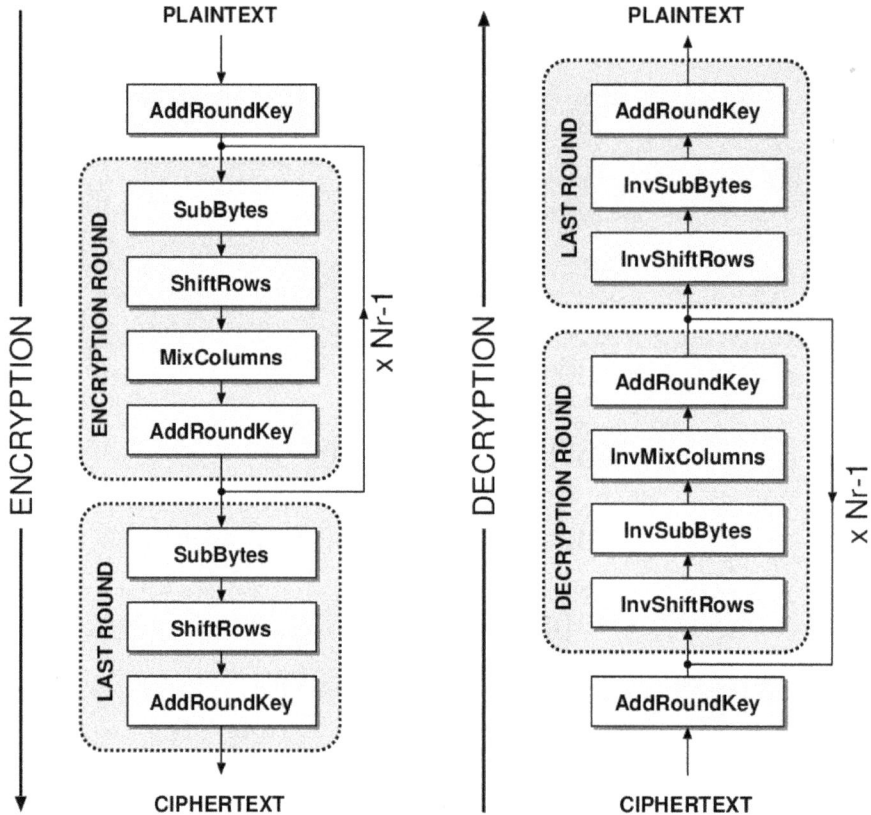

Figure 2.29 – Encryption and decryption scheme in AES

After schematizing the AES operations of encryption and decryption, we now analyze the attacks and vulnerabilities on this algorithm.

Attacks and vulnerabilities in AES

The NSA and NIST publications deemed AES as invulnerable to any kind of known attack.

However, AES has its vulnerabilities; in fact, every system that can be implemented has vulnerabilities.

Important Note

Recall the NIST document reporting the possible vulnerabilities of AES (documented on October 2, 2000). You can read it at https://www.nist.gov/news-events/news/2000/10/commerce-department-announces-winner-global-information-security.

In the NIST document, it states *Each of the candidate algorithms was required to support key sizes of 128, 192 and 256 bits. For a 128-bit key size, there are approximately 340,000,000,000,000,000,000,000,000,000,000,000,000 (340 followed by 36 zeros) possible keys.*

However, even though theoretically AES remains unbreakable, (-x%), using all the brute force in the world (you will see a computational analysis of AES in *Chapter 9, Crypto Search Engine*), it is still always possible to find a breach in any algorithm. It is common to find breaches in the implementation stage. Indeed, pay attention to what happens if you implement, for instance, AES with ECB mode. We have already seen ECB mode in DES. This basic implementation consists of dividing the plaintext into blocks and for each block of plaintext, P, calculating the ciphertext, C:

```
C = Encr (P)
```

You can see the scheme of ECB mode encryption in the following figure:

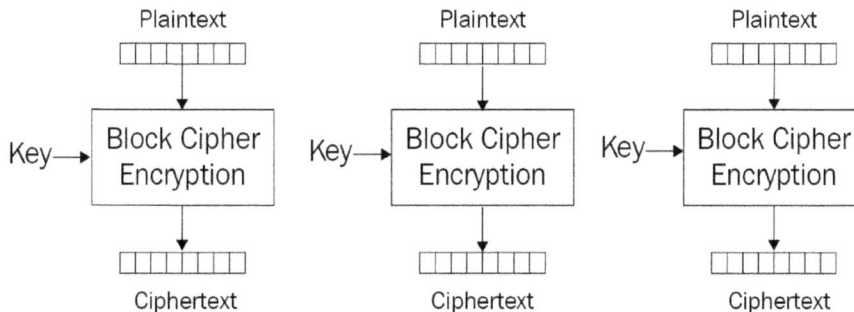

Electronic Codebook (ECB) mode encryption

Figure 2.30 – ECB mode encryption

If AES were implemented in EBC mode, as you can see in the preceding figure, in the middle (between the original and the one encrypted with another mode), there could be serious issues. For instance, in the following *ECB of Cervino mountain*, it's possible to recognize the content even though it's encrypted:

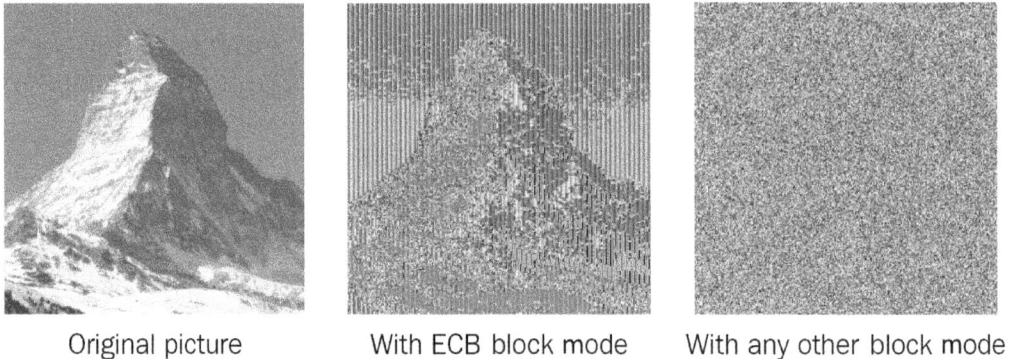

Original picture   With ECB block mode   With any other block mode

Figure 2.31 – The original picture of Cervino mountain, encrypted with ECB (failure) mode, and encrypted with another block mode

In other words, ECB block mode vanishes the encryption effect, which should have been the same as the third image (encrypted with another **block mode**).

With another attack on ECB, known as **block-reply**, knowing a plaintext-ciphertext pair, even without knowing the key, it's possible for someone to repeatedly resend the known ciphertext.

Now, an interesting example of this implication given by ECB mode is presented by *Christopher Swenson* in his book *Modern Cryptanalysis*. If Eve (the attacker) tries to trick Bob and Alice during the phase of information exchange, Eve can re-send the block of known plaintext with considerable advantage to herself.

For example, consider this hypothetical scenario related to the ECB attack mentioned previously.

Alice owns a bank account, and she goes to an ATM to withdraw money. It is assumed that the communication between Alice and the bank via the ATM is encrypted, and we suppose it would be encrypted using AES/ECB mode.

So, the encrypted message between Alice (with the key [K]) and the bank is as follows:

1. ATM: Encrypt [K] (name: Alice Smith, account number: 123456, amount: $200).

2. Let's say that this message encrypted with AES comes out in this form:

    ```
    CF   A3   1C   F4   67   T3   2D   M9......
    ```

3. Answer from the bank after having checked the account: [OK] .

If Eve is listening to the communication and intercepts the encrypted message, she could just repeat the operation several times until she steals all of Alice's money from the account. The bank would just think that Alice is making several more ATM withdrawals, and if no action is taken against this attack, the victim (Alice) will have just lost all the money in her account. This trick works because Eve resends the same message copied several times. If the [K] key doesn't change for every session of encryption, Eve can attempt this attack. A very efficient solution is to accept only different cryptograms for each session, which gets rid of the use of symmetric encryption for multiple transmissions. Otherwise, in order to prevent this kind of attack, one of the implementations of AES (just like other block ciphers) is CBC.

CBC performs the block encryption generating an output based on the values of the previous blocks.

We will see this implementation again later on, in *Chapter 9, Crypto Search Engine,* in the *Computational Analysis on CSE* section, when I present CSE in which we have implemented AES encryption in CBC mode.

Here I'll just explain how it works.

The CBC method uses a 64-bit block size for plaintext, ciphertext, and the initialization vector, IV. Essentially, IV is a random number, sometimes called *salt*, XORed to the plaintext in order to compute the block.

Just remember the following:

- E = Encryption
- D = Decryption
- C = Ciphertext

Encryption works as follows:

1. Calculate the initial block, C_0, taking the first block of the plaintext, (P_0), XORing with the initialization vector (IV):

    $$C_0 = E (P_0 \oplus IV)$$

2. Each successive block is calculated by XORing the previous ciphertext block with the plaintext block and encrypting the result:

    $$C_i = E (P_i \oplus C_{i-1})$$

Decryption works as follows:

1. To obtain the plaintext, (P_0), combine XOR between the decryption of the first block of ciphertext received (C_0) and the initialization vector (IV):

    $$P_0 = D (C_0) \oplus IV$$

2. To obtain all the other plaintext, (P_i), we have to perform an XOR between the decryption of the ciphertext received, (C_i), and all the other ciphertext excluding the first one:

    $$P_i = D (C_i) \oplus C_{i-1}$$

In the following figure, the scheme of CBC mode encryption is represented:

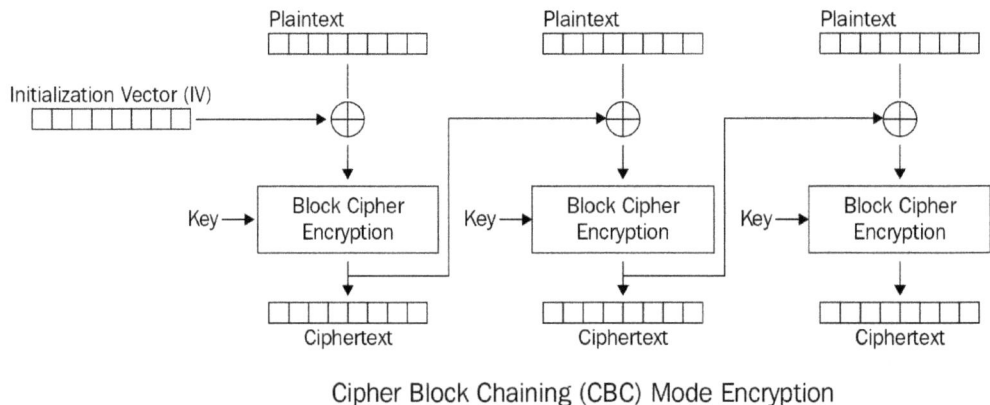

Cipher Block Chaining (CBC) Mode Encryption

Figure 2.32 – Scheme of CBC mode encryption

AES is a robust symmetric algorithm, and until 2009, the only successful attacks were so-called side-channel attacks. These kinds of attacks are mostly related to the implementation of AES in some specific applications.

Here is a list of side-channel attacks:

- **Cache attack**: Usually, some information is stored in a memory cache (a kind of memory of second order in the computer); if the attacker monitors cache access remotely, they can steal the key or the plaintext. To avoid that, it is necessary to keep the memory cache clean.

- **Timing attack**: This is a method that exploits the time to perform encryption based on the correlation between the timing and values of the parameters. If an attacker knows part of the message or part of the key, they can compare the real and modeled executed times. Essentially, it could be considered a *physical* attack on a bad implementation of the code much more than a logical attack. Anyway, this attack is not only referred to as AES but also RSA, D-H, and other algorithms too, which rely on the correlation of parameters.

- **Power monitoring attack**: Just like the timing attack, there could be potential vulnerabilities inherent to hardware implementation. You can find an interesting attack experiment at the following link relating to the correlation of the power consumption of an AES 128-bit implementation on Arduino Uno. The attack affected the ARK and SB functions of this algorithm, gaining the full 16-byte cipher key monitoring the device's power consumption. For hardware lovers, this is an exciting attack:

  `https://www.tandfonline.com/doi/full/10.1080/23742917.2016 .1231523`

- **Electromagnetic attack**: This is another kind of attack performed on the implementation of the algorithm. One attack was attempted on FPGA measuring the radiation that emanated from an antenna through an oscilloscope.

Finally, the real problem of AES is the exchanging of the key. This being a symmetric algorithm, Alice and Bob must agree on a shared key in order to perform the encryption and decryption required. Even if AES's few applications could be implemented without any key exchange, most need asymmetric algorithms to make up for the lack of key transmission. We will better understand this concept in the next chapter when we explore asymmetric encryption. Moreover, we will see an application that doesn't need any key exchange in *Chapter 9*, *Crypto Search Engine*, when analyzing the CSE.

Summary

In this chapter, you learned about symmetric encryption. We have explored the Boolean operations necessary for understanding symmetric encryption, KE, and S-Box functionality. Then, we deep-dived into how simple DES, DES, 3DES, and DESX work and their principal vulnerabilities and attacks.

After these topics, we analyzed AES (Rijndael), including its implementation schema and the logic of the steps that make this algorithm so strong. Regarding the vulnerabilities and attacks on AES, you have understood how the difference between ECB mode and CBC mode can make it vulnerable to block cipher implementation attacks.

Finally, we explored some of the best-known side-channel attacks valid for most cryptographic algorithms.

These topics are essential because now you have learned how to implement a cryptographic symmetric algorithm, and you have more familiarity with its peculiarities. We will see many correlations with this part in the next chapters. *Chapter 9*, *Crypto Search Engine*, will explain CSE, which adopts AES as one of the algorithms for the transmission of encrypted files in the cloud. Now that you have learned about the fundamentals of symmetric encryption, it's time to analyze asymmetric encryption.

3
Asymmetric Encryption

Asymmetric encryption means using different pairs of keys to encrypt and decrypt a message. A synonym of asymmetric encryption is public/private key encryption, but there are some differences between asymmetric and public/private key encryption, which we will discover in this chapter. Starting with a little bit of history of this revolutionary method of encryption/decryption, we will look at different types of asymmetric algorithms and how they help secure our credit cards, identity, and data.

In this chapter, we are going to cover the following topics:

- Public/private key and asymmetric encryption
- The Diffie-Hellman key exchange and the related man-in-the-middle problem
- RSA and an interesting application for international threats
- Introduction to conventional and unconventional attacks on RSA
- **Pretty Good Privacy (PGP)**
- ElGamal and its possible vulnerabilities

Let's dive in!

Introduction to asymmetric encryption

The most important function of private/public key encryption is exchanging a key between two parties, along with providing secure information transactions.

To fully understand asymmetric encryption, we must understand its background. This kind of cryptography is particularly important in our day-to-day lives. This is the branch of cryptography that's deputed to cover our financial secrets, such as credit cards and online accounts; to generate the passwords that we use constantly in our lives; and, in general, to share sensitive data with others securely and protect our privacy.

Let's learn a little bit about the history of this fascinating branch of cryptography.

The story of asymmetric cryptography begins in the late 1970s, but it advanced in the 1980s when the advent of the internet and the digital economy started to introduce computers to family homes. The late 1970s and 1980s was the period in which Steve Jobs founded Apple Inc. and was during the Cold War between the USA and the USSR. It was also a period of economic boom for many Western countries, such as Italy, France, and Germany. And finally, it was the period of the advent of the internet. The contraposition of the two blocs, Western and Eastern, with US allies on one side and the Soviet bloc on the other side, created opposing networks of spies that had their fulcrum in the divided city of Berlin. During this period, keys being exchanged in symmetric cryptography reached the point that the US Government's **Authority for Communications Security** (**COMSEC**), which is responsible for transmitting cryptographic keys, transported tons of keys every day. This problematic situation degenerated to a breaking point. Just to give an example, with the DES algorithm in the 1970s, banks dispatched keys via a courier that were handed over in person. The **National Security Agency** (**NSA**) in America struggled a lot with the key distribution problem, despite having access to the world's greatest computing resources. The issue of key distribution seemed to be unsolvable, even for big corporations dedicated to solving the hardest problems related to the future of the world, such as RAND – another powerful institution created to manage the problems of the future and to prevent breakpoint failures. I think that, sometimes, a breakpoint is just a way to clear up the situation instead of just ignoring it. Sometimes, issues have different ways they can be solved. In the case of asymmetric encryption, no amount of government money nor supercomputers with infinite computation and multiple brains at their service could solve a problem that, at a glance, would appear rather easy to solve.

Now that you have an idea of the main problem that asymmetric encryption solves, which is the key exchange (actually, we will see that in RSA, this problem gets translated into the direct transmission of the message), let's go deeper to explore the pioneers involved in the history of this extremely intriguing branch of cryptography.

The pioneers

Cryptographers can often appear to be strange people; sometimes, they could be introverts, sometimes extroverts. This is the case with *Whitfield Diffie*, an independent freethinker, not employed by the government or any of the big corporations. I met Diffie for the first time at a convention in San Francisco in 2016 while he was discussing cryptography with his famous colleagues, *Martin Hellmann* and *Ronald Rives*. One of the most impressive things that remained fixed in my mind was his elegant white attire, counterposed by his tall stature and long white hair and beard. Similar to an ever-young guy still in the 1960s, someone whose contemporary could be an agent at the Wall Street Stock Exchange or a holy man in India. He is one of the fathers of modern cryptography, and his name will be forever imprinted in the history of public/private key encryption. Diffie was born in 1944 and graduated from MIT in Boston in 1965. After his graduation, he was employed in the field of cybersecurity and later became one of the most authentic independent cryptographers of the 1970s. He has been described as *cyberpunk*, in honor of the *new wave* science fiction movement of the 1960s and 1970s, where cybernetics, artificial intelligence, and hacker culture combined into a dystopian futuristic setting that tended to focus on a *combination of low life and high tech*.

Back in the 1960s, the US Department of Defense began funding a new advanced program of research in the field of communication called **Defense Advanced Research Projects Agency (DARPA)**, also called **ARPA**. The main ARPA project was to connect military computers to create a more resilient grade of security in telecommunications. The project intended to prevent a blackout of communications in the event of a nuclear attack, but also, the network allowed dispatches and information to be sent between scientists, as well as calculations that had been performed, to exploit the spare capacity of the connected computers. **ARPANET** started officially in 1969 with the connection of only four sites and grew quickly so much that in 1982, the project spawned the internet. At the end of the 1980s, many regular users were connected to the internet, and thereafter their number exploded.

While ARPANET was growing, Diffie started to think that one day, everyone would have a computer, and with it, exchange emails with each other. He also imagined a world where products could be sold via the internet and real money was abandoned in favor of credit cards. His great consideration of privacy and data security led to Diffie being obsessed with the problem of how to communicate with others without having any idea who was at the opposite end of the cable. Moreover, encrypting messages and documents is often done when sending highly valuable information; data encryption was starting to be used by the general public to hide information and share secrets with others. This was the time when the usage of cryptography became common, and it was not just for the military, governments, or academics.

The main issue to solve was that if two perfect strangers meet each other via the internet, how would it be possible to encrypt/decrypt a shared document without exchanging any additional information except for the document itself, which is encrypted/decrypted through mathematical parameters? This is the key exchange problem in a nutshell.

One day in 1974, Diffie went to visit IBM's *Thomas Watson* laboratory, where he was invited to give a speech. He spoke about various strategies for attacking the key distribution problem but the people in the room were very skeptical about his solution. Only one shared his vision: he was a senior cryptographer for IBM who mentioned that a Stanford professor had recently visited the laboratory and had a similar vision about the problem. This man was Martin Hellman.

Diffie was so enthusiastic that the same evening, he drove to the opposite side of America to Palo Alto, California to visit Professor Hellman. The collaboration between Diffie and Hellman will remain famous in cryptography for the creation of one of the most beautiful algorithms in the field: the **Diffie-Hellman key exchange**.

We'll analyze this pioneering algorithm in the next section.

The Diffie-Hellman algorithm

To understand the **Diffie-Hellman (D-H)** algorithm, we can rely on the so-called *thought experiments* or *mental representation* of a theory often used by Einstein.

A thought experiment is a hypothetical scenario where you mentally transport yourself to a more real situation than in the purely theoretical way of facing an issue. For example, Einstein used a very popular thought experiment to explain the theory of relativity. He used the metaphor of a moving train observed by onlookers from different positions, inside and outside of the train.

I will often apply these mental figurative representations in this book.

Let's imagine that we have our two actors, Alice and Bob, who want to exchange a message (on paper) but the main post office in the city examines the contents of all letters. So, Alice and Bob struggled a lot with different methods to send a letter secretly while avoiding any intrusion; for example, putting a key inside a metallic cage and sending it to Bob. But because Bob wouldn't have the key to open the cage, Alice and Bob would have to meet somewhere first so that Alice could give the key to Bob. Again, we return to the problem of exchanging keys.

After many, many attempts, it seemed to be impossible to arrive at a logical solution that would solve this problem, but finally, one day, Diffie, with Hellmann's support, found the solution, about which Hellmann later said, *"God rewards fools!"*

Let's explore a mental representation of what Alice and Bob should do to exchange the key:

- **Step 1**: Alice puts her secret message inside a metallic box closed with an iron padlock and sends it to Bob, but holds on to the key herself. Now, remember that Alice locks the box using her key and doesn't give it to Bob.

- **Step 2**: Bob applies one more lock to the cage using his private key and resends the box to Alice. So, after Bob has received the box for the first time, he can't open it. He just adds one more lock to the box.

- **Step 3**: When Alice receives the box the second time, she is free to remove her padlock since the box remains secured with Bob's key, as shown in the following diagram, when Alice resends the box for the last time. Remember that the message is always inside the box. Right now, the box is only locked with Bob's padlock.

- **Step 4**: When Bob receives the box this time, he can open it, because the box only remains locked with his padlock. Finally, Bob can read the content of the message sent by Alice that has been preserved inside the box.

As you can see, Alice and Bob never met each other to exchange any padlock keys. Note that in this example, the box was sent twice from Alice to Bob, while in the algorithm, it is not:

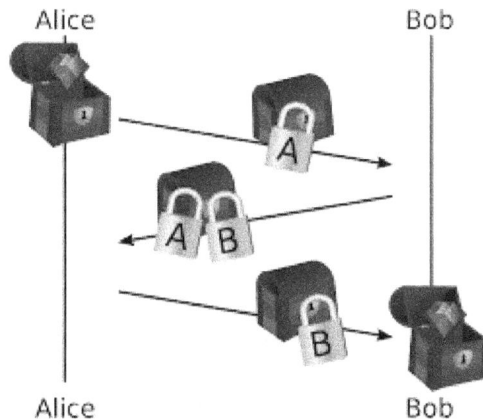

Figure 3.1 – The D-H algorithm using the Alice and Bob example

The preceding explanation and representation are not exactly what the algorithm does, but it provides a practical solution to a problem believed unsolvable: the key exchange problem.

Now, we have to transpose this practical argumentation into a logical-mathematical representation.

First of all, we will return to using modular mathematics while taking advantage of some particular properties of the operations made in finite fields.

The discrete logarithm

I will try to explain the math behind cryptography without using excessive notations, just because I don't want you to get confused, or to load this book with heavy mathematical dissertations. It's not within the scope of this book.

When we talk about finite fields, we are considering a finite group of (n) integer numbers, (Z), laying in a ring – let's say, (Zn). This group of numbers will be subjected to all the same mathematical laws, such as operations with standard integers. Since we are working in a finite field called (modulo n) here, we have to consider some critical issues that involve modular mathematics. As we saw in the previous chapter, operating in *modulus* means wrapping back to the first number each time we arrive at the end of the set. This is just like the clock's math, where we wrap back to 1 when we reach 12.

Essentially, remember that in a finite field, there is a *numerical period* in which the numbers and the results of the operations of the field recur. For example, if we have a set of seven integers, {0, 1, 2, 3, 4, 5, 6}, often abbreviated as (Z7), we have all the operations that have been performed inside this finite field wrapping back inside the integers of the field.

Here is a short example of operations within a finite field, (Z7, +, x), of addition and multiplication. Since all the operations, (modulo 7), will work, we have to consider that the numbers will wrap back to 0 each time the operation exceeds the number 7:

```
1 x 1 ≡ 1 (mod 7)
2 x 4 ≡ 1 (mod 7)
3 + 5 ≡ 1 (mod 7)
3 x 5 ≡ 1 (mod 7)
```

Therefore, let's use this modality of counting and consider that the = notation is equivalent to ≡, which is the mathematics we learned at elementary school, where 2 + 2 = 4 doesn't properly work if we consider, for example, a finite field of modulo 3:

```
2 + 2 ≡ 1 (mod 3)
```

From high school mathematics, we recall that the [log a(z)] logarithm is a function where (a) is the base. We have to determine the exponent to give to (a) to obtain the number (z). So, for example, if a = 10 and z = 100, we find that the logarithm is 2 and we say that logarithm base 10 of 100 is 2. If we use Mathematica to calculate a logarithm, we have to compose a different notation, that is, Log[10, 100] = 2.

While working in the discrete field, things became more complicated, so instead of using a normal logarithm, we started working with a discrete logarithm.

So, let's say we have to solve an equation like this:

```
a^[x] ≡ b (mod p)
```

This would be a very hard problem, even if we know the value of (a) and (b), because there is no efficient algorithm known to solve the discrete logarithm, that is, [x].

Important Note

I have used square brackets to say that [x] is secret. Technically [x] is a discrete power, but the problems of searching for discrete logarithm and discrete power have the same computational difficulty.

Let's go a little bit deeper now to explain the dynamics of this operation. Let's consider computing the following:

```
2^4 ≡ (x) (mod 13)
```

First, we calculate 2^4 = 16, then divide 16:13 = 1 by the remainder of 3 so that x = 3.

The discrete logarithm is just the inverse operation:

```
2^[y] ≡ (x) (mod 13)
```

In this case, we have to calculate [y] while knowing the base is 2. It's possible to demonstrate that there are infinite solutions for [y] that generate (x).

Taking the preceding example, we have the following:

```
2^[y] ≡ 3 (mod 13) for [y]
```

One solution is y = 4, but it is not the only one.

The result of 3 is also satisfied for all the integers, (n), of this equation:

```
[y] = [y + (p-1)*n]
```

Let's prove n = 1:

```
2^[4+(13-1)*1] ≡ 2^16 (mod 13)
2^16 ≡ 3 (mod 13)
```

But it is also valid for n = 2:

```
2^[4 +(13-1)*2] ≡ 2^28 (mod 13)
2^28 ≡ 3 (mod 13)
```

And so on...

Hence, the equation has infinite solutions for all the integers; that is, (n ≥ 0):

```
[y] ≡ 2^[4 + 12 n] (mod 13)
```

There is no method yet for solving the discrete logarithm in polynomial time. So, in mathematics, as in cryptography, this problem is considered very hard to solve, even if the attacker has a lot of computation power.

Finally, we have to introduce the definition and the notation of a generator, (g), which is a particular number where we say that (g) generates the entire group, (Zp). If (p) is a prime number, this means that (g) can take on any value between 1 and p-1.

Explaining the D-H algorithm

D-H is not exactly an asymmetric encryption algorithm, but it can be defined properly as a public/private key algorithm. The difference between asymmetric and public/private keys is not only formal but substantial. You will understand the difference better later, in the *RSA* section, which covers a pure asymmetric algorithm. Instead, D-H gets a shared key, which works to symmetrically encrypt the message, [M].

The encryption that's performed with a symmetric algorithm is combined with a D-H shared key transmission to generate the cryptogram, C:

```
Symmetric-Algorithm E((D-H[k]), M) = C
```

In other words, we use the D-H algorithm to generate the shared secret key, [k], then with AES or another symmetric algorithm, we encrypt the message, [M].

D-H doesn't directly encrypt the secret message; it can only determine a shared key between two parties. This is a critical point, as we will see in the next paragraph.

However, for working in discrete fields and applying a discrete logarithm problem to shield the key from attackers when sharing it, Diffie and Hellman implemented one of the most robust and famous algorithms in cryptography.

Let's see how D-H works:

- **Step 1**: Alice and Bob first agree on the parameters: (g) as a generator in the ring, (Zp), and a prime number, p (mod p).

  Step 2: Alice chooses a secret number, [a], and Bob chooses a secret number, [b]. Alice calculates A ≡ g^a (mod p) and Bob calculates B ≡ g^b (mod p).

- **Step 3**: Alice sends (A) to Bob and Bob sends (B) to Alice.

- **Step 4**: Alice computes ka ≡ B^a (mod p) and Bob computes kb ≡ A^b (mod p).

  [ka = kb] will be the secret that's shared key between Alice and Bob.

The following is a numerical example of this algorithm:

- **Step 1**: Alice and Bob agree on the parameters they will use; that is, g = 7 and (mod 11).

  Step 2: Alice chooses a secret number, (3), and Bob chooses a secret number, (6). Alice calculates 7^3 (mod 11) ≡ 2 and Bob calculates 7^6 (mod 11) ≡ 4.

- **Step 3**: Alice sends (2) to Bob and Bob sends (4) to Alice.

- **Step 4**: Alice computes 4^3 (mod 11) ≡ 9 and Bob computes 2^6 (mod 11) ≡ 9.

The number [9] is the shared secret key, [k], of Alice and Bob.

Analyzing the algorithm

In *Step 2* of the algorithm, Alice calculates A ≡ g^a (mod p) and Bob calculates B ≡ g^b (mod p). Alice and Bob have exchanged (A) and (B), the public parameters of the one-way function. A *one-way function* has this name because it is impossible, (-x%), to return from the public parameter, (A), to calculate the secret private key, [a] (and the same for (B) with [b]), for the robustness of the discrete logarithm (see the *The discrete logarithm* section).

Another property of the modular powers is that we can write B^a and A^b (mod p), as follows:

```
B ^a ≡ (g^b)^a (mod p)
A ^b ≡ (g^a)^b (mod p)
```

So, for the property of modular exponentiations, we have the following:

```
g^(b*a) ≡ g^(a*b) (mod p)
```

For example, we have the following:

- Alice: `(7^6)^3 ≡ 7^(6*3) ≡ 9 (mod 11)`
- Bob: `(7^3)^6 ≡ 7^(3*6) ≡ 9 (mod 11)`

This is the mathematical trick that makes it possible for the D-H algorithm to work.

Now that we have understood the algorithm, let's highlight its defects and the possible attacks that can be performed on it.

Possible attacks and cryptanalysis on the D-H algorithm

The most common attack that's performed on the D-H algorithm is the **man-in-the-middle (MiM)** attack.

A MiM attack is when the attacker infiltrates a channel of communication and spies on, blocks, or alters the communication between the sender and the receiver. Usually, the attack is accomplished by Eve (the attacker) pretending to be one of the two true actors in the conversation:

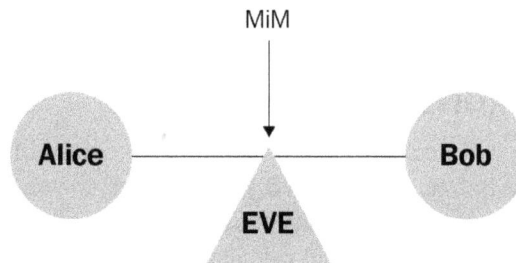

Figure 3.2 – Eve is the man in the middle

Recalling what happened in *Step 3* of the D-H algorithm, Alice and Bob exchanged their public parameters, (A) and (B).

Here, Alice sends (A) to Bob, then Bob sends (B) to Alice.

Now, Eve (the attacker) interferes within the communication by pretending to be Alice.

So, a MiM attack looks like this:

- **Step 3**: Alice sends (A) to Bob and Bob sends (B) to Alice.

- **Here, we have the MiM attack**: Eve sends (E) to Bob and then Bob sends (B) to Eve, assuming it is Alice.

Let's analyze Alice's function, `A ≡ g^a (mod p)` and Bob's function `B ≡ g^b (mod p)`. This passage would be crucial if it was done in normal arithmetic and not in finite fields.

There is another possible attack, also known as the **birthday attack**, which is one of the most famous attacks performed on discrete logarithms. It is consolidated that among a group of people, at least two of them will share a birthday so that in a cyclic group, it will be possible to find some equal values (collisions) to solve a discrete logarithm.

> **Important Note**
>
> (E) here represents Eve's public parameter, which has been generated by her private key, not encryption.

Proceeding with the final part of the algorithm, you can see the effects of the MiM trick.

Suppose that Alice and Bob are using D-H to generate a shared private key to encrypt the following message:

Alice's message: *Bob, please transfer $10,000.00 to my account number 1234567.*

After *Step 3*, in which Bob and Eve (pretending to be Alice) have exchanged their public parameters, Eve sends the modified message to Bob (intercepted from Alice), which has been encrypted with the shared key.

Suppose, the encrypted message from Eve is `bu3fb3440r3nrunfjr3umi4gj57*je`.

Bob receives the preceding encrypted message (supposedly from Alice) and decrypts it with the D-H shared key, obtaining some plaintext.

Eve's modified message after the MiM attack is *Bob, please transfer $10,000.00 to my account number 3217654.*

As you will have noticed, the account number is Eve's account. This attack is potentially disruptive.

Analyzing the attack, *Step 3* is not critical (as we have said) because (A) and (B) have been communicated in clear mode, but the question is: how can Bob be sure that (A) is coming from Alice?

The answer is, by using the D-H algorithm, Bob can't be sure that (A) comes from Alice and not from Eve (the attacker); similarly, Alice can't know that (B) comes from Bob either. In the absence of additional information about the identity of the two parties, relying only on the parameters received, the D-H algorithm suffers from this possible substitution-of-identity attack called MiM

This example shows the need for the sender (Alice) and the receiver (Bob) to have a way to be sure that they are who they say they are, and that their public keys, (A) and (B), do come from Alice and Bob, respectively. To prevent the problem of a MiM attack and identify the users of the communication channel, one of the most widely used techniques is a digital signature. We will look at digital signatures in *Chapter 4*, *Introducing Hash Functions and Digital Signatures*, which is entirely dedicated to explaining these cryptographic methods.

Moreover, it's possible for a public/private algorithm to identify the parties and overcome the MiM attack. In *Part 4* of this book, I will show you some public/private key algorithms of the new generation that, although not asymmetric, have multiple ways to be signed.

Finally, a version of D-H can be implemented using elliptic curves. We will analyze this algorithm in *Chapter 7*, *Elliptic Curves*.

RSA

Among the cryptography algorithms, RSA shines like a star. Its beauty is equal to its logical simplicity and hidden inside is such a force that after 40 years, it's still used to protect more than 80 percent of commercial transactions in the world.

Its acronym is made up of the names of its three inventors – **Rivest, Shamir, and Aldemar. RSA** is what we call the perfect asymmetric algorithm. Actually, in 1997, the CESG, an English cryptography agency, attributed the invention of public-key encryption to James Allis in 1970 and the same agency declared that in 1973, a document was written by Clifford Cocks that demonstrates a similar version of the RSA algorithm.

The essential concept of the asymmetric algorithm is that the keys for encryption and decryption are different.

Recalling the analogy to padlocks I made in the *The Diffie-Hellman algorithm* section, when I described the D-H algorithm, we saw that anybody (not just Alice and Bob) could lock the box with a padlock. This is the true problem of MiM because the padlock can't be recognized as specifically belonging to Bob or Alice.

To overcome this problem, another interesting mental experiment can be done using a similar analogy but making things a little different.

Suppose that Alice makes many copies of her padlock, and she sends these copies to every postal office in the country, keeping the key that opens all the padlocks in a secret place.

If Bob wishes to send a secret message to Alice, he can go to a postal office and ask for Alice's padlock, put the message inside the box, and lock it.

In this way, Bob (the sender), from the moment he locks the box, will be unable to unlock it, but when Alice receives the box, she will be able to open it with her unique secure key.

Basically, in RSA (as opposed to D-H), Bob encrypts the message with Alice's public key. After the encryption process, even Bob is unable to decrypt the message, while Alice can decrypt it using her private key.

This was the step that transformed the concept of asymmetric encryption from mere theory to practical use. RSA discovered how to encrypt a message with the public key of the receiver and decrypt it with the private key. To make that possible, RSA needs a particular mathematic function that I will explain further later on when we explore the algorithm in detail.

As we mentioned previously, there were three inventors of this algorithm. They were all researchers at MIT, Boston, at the time. We are talking about the late 1970s. After the invention of the D-H algorithm, Ronald Rivest was extremely fascinated by this new kind of cryptography. He first involved a mathematician, Leonard Adleman, and then another colleague from the Computer Science department, Aid Shamir, who joined the group. What Rivest was trying to achieve was a mathematical way to send a secret message encrypted with a public key and decrypt it with the private key of the receiver. However, in D-H, the message can only be encrypted once the key has been exchanged, using the same shared key. Here, the problem was to find a way to send the message that had been encrypted with a public key and decrypted through the private key. But as I've said, it needed a very particular inverse mathematical function. This is the real added value of the RSA invention that we are going to discover shortly.

The tale of this discovery, as Rivest told it, is funny. It was April 1977 when Rivest and Adleman met at the home of a student for Easter. They drank a little too much wine and at about midnight, Rivest went back home. He started to think over the problem that had been tormenting him for almost a year. Laying on his bed, he opened a mathematics book and discovered the function that could be perfect for the goal that the group had.

The function he found was a particular inverse function in modular mathematics related to the factorization problem.

As I introduced in *Chapter 1, Deep Diving into Cryptography*, the problem of factorizing a large number made by multiplying two big prime numbers is considered very hard to solve, even for a computer with immense computation power.

Explaining RSA

To understand this algorithm, we will consider Alice and Bob exchanging a secret message.

Let's say that Bob wants to send a secret message to Alice, given the following:

- M: The secret message

- e: A public parameter (usually a fixed number)

- c: The ciphertext or cryptogram

- p, q: Two large random prime numbers (the private keys of Alice)

The following is the public and private key generation. As you will see, the core of RSA (its magic function) is generating Alice's private key, [d].

Key generation:

Alice's public key, (N), is given by the following code:

```
N = p*q
```

As we mentioned earlier, multiplying two big prime numbers makes (N) very difficult to factorize, and makes it generally very difficult for an attacker to find [p] and [q]. Alice's private key, [d], is given by the following code:

```
[d] * e ≡ 1 (mod[p-1]*[q-1])
```

Bob performs the encryption:

```
c ≡ M^e (mod N)
```

Bob sends the ciphertext, (c), to Alice. She can now decrypt (c) using her private key, [d]:

```
c^d ≡ M (mod N)
```

And that's it!

> **Note**
>
> The bold elements – M, d, p, and q – in the algorithm are protected and secret.

Numerical example:

```
M = 88
e = 9007
p = 101
q = 67
N = 6767
```

Step 1: Bob's encryption is as follows:

```
88^9007 ≡ 6621 (mod 6767)
```

Alice receives a cryptogram, that is, `c = 6621`.

Step 2: Alice's decryption is as follows:

```
9007* d ≡ 1 (mod (101-1)*(67-1))
d = 3943
6621^3943 ≡ 88 (mod 6767)
```

As you can see, the secret message, `[M] = 88`, comes back from Alice's private key, `[d] = 3943`.

Analyzing RSA

There are several elements to explain but the most important is to understand why this function, which is used for decrypting `(c)` and obtaining `[M]`, works:

```
M ≡ c^[d] (mod N)
```

This is *Step 2*; that is, the decryption function. I have just inverted the notation by putting `[M]` on the left.

The reason it works is hidden in the key generation equation:

```
[d] * e ≡ 1 (mod [p-1]*[q-1])
```

`[d]` is Alice's private key. For Euler's theorem, the function will probably be verified because the numbers `[p]` and `[q]` are very big and `[M]` is probably a co-prime of `(N)`. If this equation is verified, then we can rewrite the encryption stage as follows:

```
(M^e)^d (mod N)
```

For the properties of the powers and Euler's theorem, we have the following:

```
M^(e*d)  (mod N)
de ≡ 1  (mod (p-1)*(q-1))
```

That is the same as writing `M^1 = M (mod N)`.

So, by inserting `[d]` inside the decryption stage, Alice can obtain `[M]`.

Conventional attacks on the algorithm

All the attacks that will be explained in the first part of this section are recognized and well known. That is why we are talking about conventional attacks on RSA.

The first three methods of attack on RSA are related to the `(mod N)` public parameter. To perform an attack on `N = p*q`, the attacker could do the following:

- Use an *efficient* algorithm of factorization to discover p and q.

- Use new algorithms that, under certain conditions, can find the numbers.

- Use a quantum computer to factorize `N` (in the future).

Let's analyze the following three cases:

- In the first case, an efficient algorithm of factorization is not yet known. The most common methods are as follows:

  - The general number field sieve algorithm

  - The quadratic sieve algorithm

  - The Pollard algorithm

- In the second case, if (n) is the number of digits of `N = p*q` and the attacker knows the first (n/4) digits or the last (n/4) digits of `[p]` or `[q]`, then it will be possible to factorize (N) in an efficient way. Anyway, there is a very remote possibility of knowing it. For example, if `[p]` and `[q]` have 100 digits and the first (or the last) 50 digits of `[p]` are known, then it's possible to factorize N.

For more information, you can refer to Coppersmith's method of factorization. More cases related to Coppersmith attacks, as explained later in this section, are where the exponents, (e) or [d], and even the plaintext, [M], are too short.

- If an attacker uses a quantum computer, it will be theoretically possible to factorize N in a short time with Shor's algorithm, and I am convinced that in the future, other, more efficient quantum algorithms will arise. I will explain this theory in more detail in *Chapter 8, Quantum Cryptography*, where we talk about quantum computing and Q-cryptography.

Finally, if we have a very short piece of plaintext, [M], and even the exponent, (e), is short, then RSA could be breakable. This is because the power operation, M^e, remains inside modulo N. So, in this phase of encryption, let's say we have the following:

```
M^e < N
```

Here, it's enough to use the e-th root of (c) to find]M[.

> **Important Note**
> I have used open brackets,]M[, to denote that the message has been decrypted.

Numerical example:

```
M = 2
e = 3
N = 77
2^3 ≡ 8  (mod 77)
```

Since e = 3, by performing a simple cubic root, √, we can obtain the message in cleartext:

```
8^(1/3) = 2
```

Here, we are working in linear mathematics and no longer in modular mathematics.

A way to overcome this problem is to lengthen the message by adding random bits to it. This method is very common in cryptography and cybersecurity and is known as **padding**.

There are different ways to perform padding, but here, we are talking about **bit padding**. As we covered in *Chapter 1, Deep Diving into Cryptography*, we can use ASCII code to convert text into a binary system, so the message, [M], is a string of bits. If we add random bits (usually at the end, though they could also be added at the start), we will obtain something like this:

```
... | 1011 1001 1101 0100 0010 0111 0000 0000 |
```

As you can see, the bold digits represent the padding.

This method can be used to pad messages that are any number of bits long, not necessarily a whole number of bytes long; for example, a message of 23 bits that is padded with 9 bits to fill a 32-bit block.

Now that we are more familiar with RSA and modular mathematics properties, we'll explore the first interesting application that was implemented with this algorithm.

The application of RSA to verify international treaties

Let's say that Nation Alpha wants to monitor seismic data from Nation Beta to be sure that they don't experiment with any nuclear bombs in their territory. A set of sensors has been installed on the ground of Nation Beta to monitor its seismic activity, recorded, and encrypted. Then, the output data is transmitted to Nation Alpha, let's say, via a satellite.

This interesting application of RSA works as follows:

- Nation Alpha, (A), wants to be sure that Nation Beta, (B), doesn't modify the data.
- Nation Beta wants to check the message before sending it (for spying purposes).

We name the data that's collected from the sensors [x]; so, the protocol works as follows:

Key generation:

1. Alpha chooses the parameters, (N= p*q), as the product of two big prime numbers, and the (e) parameter.
2. Alpha sends (N, e) to Beta.
3. Alpha keeps the private key, [d], secret.

The *protocol* for threat verification on atomic experiments is developed as follows:

- **Step 1**: A sensor located deep in the earth collects data, [x], performing encryption using the private key, [d]:

    x^d ≡ y (mod N)

- **Step 2**: Initially, both the (x) and (y) parameters are sent by the sensor to Beta to let them verify the truthfulness of the information. Beta checks the following:

    y^e ≡ x (mod N)

- **Step 3**: After the positive check, Beta forwards (x, y) to Alpha, who can control the result of (x):

    x ≡ y^e (mod N)

Important Note

(x) is the collected set of data from the sensor, while [d] is the private key of Alpha stored inside the protected software sensor that collects the data.

This encryption is performed in the opposite way to how RSA usually works.

If the y^e ≡ x (mod N) equation is verified, Alpha can be confident that the data that's been sent from Beta is correct and they didn't modify the message or manipulate the sensor. That's because the encrypted message, (x), corresponding to the cryptogram, (y), can truly be generated by only those who know the private key, [d].

If Beta has previously attempted to manipulate the encryption inside the box that holds the sensor by changing the value of (x), then it will be very difficult for Beta to get a meaningful message.

As we mentioned previously, in this protocol, RSA is inverted, and the encryption is performed with the private key, [d], instead of (e), the public parameter, which is what normally happens.

Essentially, the difficulty here for Beta in modifying the encryption is to get a meaningful number or message. Trying to modify the cryptogram, (y), even if the (x) parameter is previously known, has the same complexity to perform the discrete logarithm (which is a hard problem to solve, as we have already seen).

We visualize the process by referring to the following diagram:

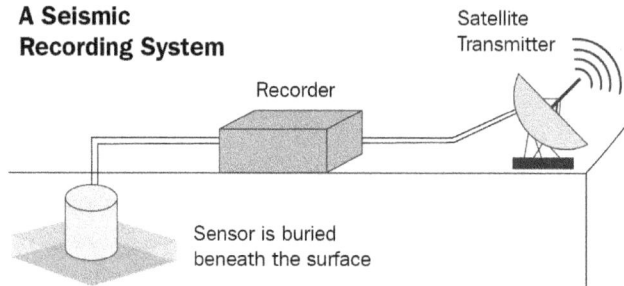

Figure 3.3 – The sensor buried underground with the data transmitted via satellite

Now that we've learned how to use the international treaties that are performed with the RSA algorithm, I want to introduce a section that will be discussed later in *Chapter 6, New Algorithms in Public/Private Key Encryption*, related to unconventional attacks on the RSA algorithm and its most famous library, OpenSSL.

Unconventional attacks

I have called these algorithms *unconventional* because they have been implemented by me and not tested and published until now.

We will see as we continue through this book, these *unconventional attacks* against RSA are even valid for other asymmetric encryption algorithms. These unconventional attacks, implemented between 2011 and 2014, have the scope to recover the secret message, [M], without knowing Alice's decryption key, [d], and the prime secret numbers, [p] and [q], behind (N). I will showcase these unconventional attacks here in this section, but I will present these methods in more detail in *Chapter 6, New Algorithms in Public/Private Key Encryption*.

Among these algorithms, we have some attacks that are used against RSA, but they can be transposed to most of the asymmetric algorithms covered in this chapter.

A new algorithm that could break the factorization problem in the future is *NextPrime*. It derives from a *genetic algorithm* discovered by a personal dear friend who explained the mechanism to me in 2009, Gerardo Iovane. In his article, *The Set of Prime Numbers*, Gerardo Iovane describes how it is possible to get all the prime numbers starting from a simple algorithm, discarding the non-prime numbers from the pattern.

Note

For a more accurate and thorough discussion of Gerardo Iovane's genetic algorithm, you can read *The Set of Prime Numbers* at https://arxiv. org/abs/0709.1539.

After years of work and many headaches, I have arrived at a mathematical function that represents a curve; each position on this curve represents a prime number, and between many positions lie the semiprimes (N) generated by two primes. This curve geometrically represents all the primes of the universe. It turns out that the position of (N) lies always in-between the positions on the curve of the two prime numbers, [p] and [q], and (N) is almost equidistant from their positions. It's also possible to demonstrate that the prime numbers have a very clear order and are not randomly positioned and disordered as believed.

The *distance* between the two prime objects, [p] and [q], of the multiplication that determines (N) is equivalent to the number of primes lying between [p] and [q]. For example, the *distance prime* between 17 and 19 is zero, the distance between 1 and 100 is 25, and the distance between 10,000 and 10,500 is only 55.

Right now, this algorithm is only efficient under determinate conditions: for example, when [p] and [q] are *rather close* to each other (at a polynomial distance). However, the interesting thing is that it doesn't matter how big the two primes are. I did some tests with this algorithm using primes of the caliber of 10^1,000 digits.

Just to clarify how big such numbers are, you can consider that 10^80 represents the number of particles in the Universe. For instance, a semi-prime with an order of 10^1,000 digits corresponds to RSA's public key length of around 3,000 bits (one of the biggest public keys that's used in cryptography right now). It could be processed in an elapsed time of a few seconds using the NextPrime algorithm if the two primes are close to each other.

At the time of writing this book, I am working on a version of the NextPrime algorithm based on a quantum computer. It could be the next generation of quantum computing factorization algorithms (similar to the Shor algorithm, which we will look at in *Chapter 8, Quantum Cryptography*).

Now, let's continue to analyze how else it is possible to attack RSA. As shown in the following diagram, there are two points of attack in the algorithm: one is the factorization of (N), while the other is the discrete power, [M^e]:

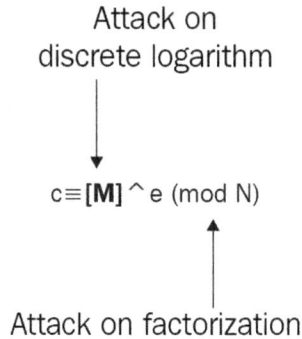

Attack on
discrete logarithm

$c \equiv$ **[M]** $^\wedge$ e (mod N)

Attack on factorization

Figure 3.4 – Points where it is possible to attack RSA

Most of the *conventional* analysis of cryptologists is focused on factorizing (N), as we have seen. But RSA suffers not only from the factorization problem; there is also another problem linked to the exponent, (e), which we will see in *Chapter 6, New Algorithms in Public/Private Key Cryptography*. These methods are essentially *backdoors* that can recover the message, [M], without knowing the secret parameters of the sender: [d], [p], and [q]. What we will understand in *Chapter 6, New Algorithms in Public/Private Key Cryptography*, when we analyze these methods of attack, is that they are equivalent to creating a backdoor inside the RSA algorithm and its main library, OpenSSL.

The RSA paradigm that we've already examined states that Bob (the sender) cannot return the message once it has been delivered. However, if we apply those unconventional methods to break RSA, this paradigm is no longer valid. Bob can create his "fake encryption" by himself to return the message, [M], encrypted with Alice's public key, while the fake cryptogram can be decrypted by Alice using RSA's decryption stage.

Is this method an unreasonable model, unrepresentative of practical situations, or does it have practical uses?

I will leave the answer to this for *Chapter 6, New Algorithms in Public/Private Key Cryptography*. Now, let's explore another protocol based on the RSA implementation that has become a popular piece of software: PGP.

PGP

Pretty Good Privacy (**PGP**) is probably the most used cryptographic software in the world.

PGP was implemented by Philip Zimmermann during the Cold War. Philip started planning to take his family to New Zealand because he believed that, in the event of a nuclear attack, the country, so isolated from the rest of the world, would be less impacted by atomic devastation. At some point, while planning to move to New Zealand, something changed his mind and he decided to remain in the US.

To communicate with his friends, Zimmermann, who was an anti-nuclear activist, developed PGP to secure messages and files transmitted via the internet. He released the software as open source, free of charge for non-commercial use.

At the time, cryptosystems larger than 40 bits were considered to be like munitions. That is to say that cryptography is still considered a military weapon. There is a requirement that you must obtain if you decide to patent a new cryptosystem, you must have authorization from the Department of Defense to publish it. This was a problem that was encountered by PGP, which never used keys smaller than 128 bits. Penalties and criminal prosecutions for violating this legal requirement are very severe, which is why Zimmermann had a *pending legal status* for many years until the American government decided to close the investigation and clear him.

PGP is not a proper algorithm, but a protocol. The innovation here is just to merge asymmetric with symmetric encryption. The protocol uses an asymmetric encryption algorithm to exchange the key, and symmetric encryption to encrypt the message and obtain the ciphertext. Moreover, a digital signature is required to identify the user and avoid a MiM attack.

The following are the steps of the protocol:

- **Step 1**: The key is transmitted using an asymmetric encryption algorithm (ElGamal, RSA).

- **Step 2**: The key that's been transmitted with asymmetric encryption becomes the session key for the symmetric encryption (DES, Triple DES, IDEA, and AES – we covered this in *Chapter 2, Introduction to Symmetric Encryption*).

- **Step 3**: A digital signature is used to identify the users (we will look at RSA digital signatures in *Chapter 4, Introducing Hash Functions and Digital Signatures*).

- **Step 4**: Decryption is performed using the symmetric key.

PGP is a good protocol for very good privacy and for securing the transmission of commercial secrets.

The ElGamal algorithm

This algorithm is an asymmetric version of the D-H algorithm. ElGamal aims to overcome the problems of MiM and the impossibility of the signatures for key ownership in D-H. Moreover, ElGamal (just like RSA) is an authentic asymmetric algorithm because it encrypts the message without previously exchanging the key.

The difficulty here is commonly related to solving the discrete logarithm. As we will see later, there is also a problem related to factorization.

ElGamal is the first algorithm we'll explore that presents a new element: an integer random number, [k], that's chosen by the sender and kept secret. It's an important innovative element because it makes its encryption "ephemeral," in the sense that [k] makes the encryption function unpredictable. Moreover, we will frequently see this new element related to the zero-knowledge protocol in *Chapter 5, Introduction to Zero-Knowledge Protocols*.

Let's look at the implementation of this algorithm and how it is used to transmit the secret message, [M].

Alice and Bob are always the two actors. Alice is the sender and Bob is the receiver.

The following diagram shows the workflow of the ElGamal algorithm:

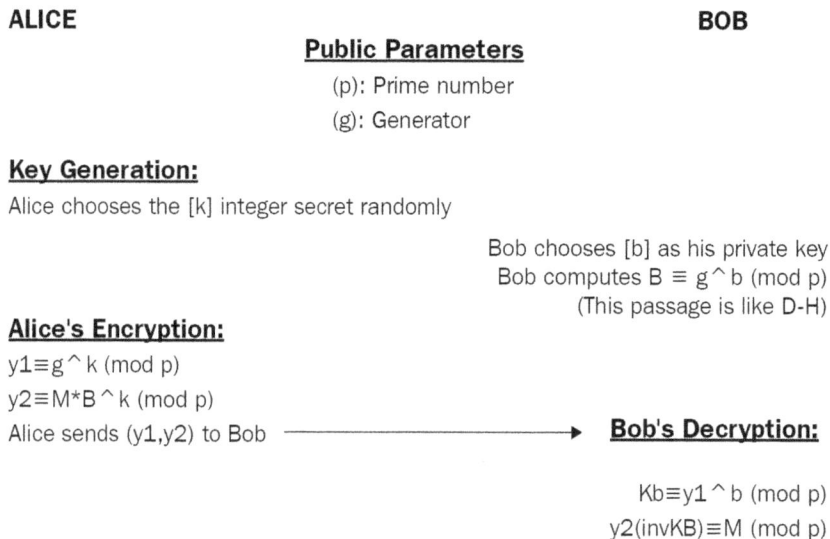

ALICE   **BOB**

Public Parameters

(p): Prime number

(g): Generator

Key Generation:

Alice chooses the [k] integer secret randomly

Bob chooses [b] as his private key

Bob computes $B \equiv g \char`^ b \pmod p$

(This passage is like D-H)

Alice's Encryption:

$y1 \equiv g \char`^ k \pmod p$

$y2 \equiv M * B \char`^ k \pmod p$

Alice sends (y1,y2) to Bob $\longrightarrow$ **Bob's Decryption:**

$Kb \equiv y1 \char`^ b \pmod p$

$y2(invKB) \equiv M \pmod p$

Figure 3.5 – Encryption/decryption of the ElGamal algorithm

As shown in the last step of Bob's decryption, we can see an inverse multiplication in (mod p). This kind of operation is essentially a division that's performed in a finite field. So, if the inverse of A is B, we have A*B = 1 (mod p). The following example shows the implementation of this inverted modular function with Mathematica.

Now, having explained the algorithm, let's look at a numerical example to understand it.

Publicly defined parameters:

The public parameters are p (a large prime number) and g (a generator):

```
p = 200003
g = 7
```

Key generation:

Alice chooses a random number, [k], and keeps it secret.

k = 23 (Alice's private key).

Bob computes his public key, (B), starting from his private key, [b]:

b = 2367 (Bob's private key):

```
B ≡ 7^2367 (mod 200003)
B = 151854
```

Alice's encryption:

Alice generates a secret message, [M]:

```
M = 88
```

Then, Alice computes (y1, y2), the two public parameters that she will send to Bob:

```
y1 ≡ 7^23 (mod 200003)
y1 = 90914
y2 ≡ 88 * 151854^ 23 (mod 200003)
y2 = 161212
```

Alice sends the parameters to Bob (y1 = 90914; y2 = 161212).

Bob's decryption:

First, Bob computes (`Kb`) by taking `y1` `=` `90914`, which is elevated to his private key, `[b]` `=` `2367`:

```
Kb ≡ 90914^2367 (mod 200003)
Kb = 10923
```

Inverted `Kb` in (`mod p`) `[Reduce[Kb*x == 1, x, Modulus -> p]` (performed with Wolfram Mathematica):

```
Inverted Kb ≡ 192331 (mod 200003)
```

Finally, Bob can return the message, `[M]`.

Bob takes (`y2`) and multiplies it by `[Inverted Kb]`, returning the message, `[M]`:

```
y2* InvKb ≡ M (mod 200003)
```

The final result is the message `[M]`:

```
161212 * 192331 ≡ 88 (mod 200003).
```

ElGamal encryption is used in the free **GNU Privacy Guard (GnuPG)** software. Over the years, GnuPG has gained wide popularity and become the de facto standard free software for private communication and digital signatures. GnuPG uses the most recent version of PGP to exchange cryptographic keys. For more information, you can go to the web page of this software: `https://gnupg.org/software/index.html`.

Important Note

As I have mentioned previously, it's assumed that the underlying problem behind the ElGamal algorithm is the discrete logarithm. This is because, as we have seen, the public parameters and keys are all defined by equations that rely on discrete logarithms.

For example, `B ≡ g^b (mod p)`; `Y1 ≡ g^k (mod p)` and `Kb ≡ y1^b (mod p)` are functions related to the discrete logarithm.

Although the discrete logarithm problem is considered to be the main problem in ElGamal, we also have the factorization problem, as shown here. Let's go back to the encryption function, (`y2`):

```
y2 ≡ M*B^k (mod p)
```

Here, you can see that there is multiplication. So, an attacker could also try to arrive at the message by factorizing (y2).

This will be clearer if we reduce the function:

```
H ≡ B^k (mod p)
```

Then, we have the following:

```
y2 ≡ M* H (mod p)
```

This can be rewritten like so:

```
M ≡ y2/H (mod p)
```

As you can see, (y2) is the product of [M*H]. If someone can find the factors of (y2), they can probably find [M].

Summary

In this chapter, we analyzed some fundamental topics surrounding asymmetric encryption. In particular, we learned how the discrete logarithm works, as well as how some of the most famous algorithms in asymmetric encryption, such as Diffie-Hellmann, RSA, and ElGamal, work. We also explored an interesting application of RSA related to exchanging sensitive data between two nations. In *Chapter 4*, *Introducing Hash Functions and Digital Signatures*, we will learn how to digitally sign these algorithms.

Now that we have learned about the fundamentals of asymmetric encryption, it's time to analyze digital signatures. As you have already seen with PGP, all these topics are very much related to each other.

4
Introducing Hash Functions and Digital Signatures

Since time immemorial, most contracts, meaning any kind of agreement between people or groups, have been written on paper and signed manually using a particular signature at the end of the document to authenticate the signatory. This was possible because, physically, the signatories were in the same place at the moment of signing. The signatories could usually trust each other because a third trustable person (a notary or legal entity) guaranteed their identities as a *super-party entity*.

Nowadays, people wanting to sign contracts often don't know each other and frequently share documents to be signed via email, signing them without a trustable third party to guarantee their identities.

Imagine that you are signing a contract with a third party and will be sending it via the internet. Now, consider the third party as *untrustable*, and you don't want to expose the document's contents to an unknown person via an unsecured channel such as the internet. How is it possible in this case to verify whether the signatures are correct and acceptable?

Moreover, how is it possible to hide the document's content and, at the same time, allow a signature on the document?

Digital signatures come to our aid to make this possible. This chapter will also show how hash functions are very useful for digitally signing encrypted documents so that anyone can identify the signers, and at the same time, the document is not exposed to prying eyes.

In this chapter, we will cover the following topics:

- Hash functions
- Digital signatures with RSA and El Gamal
- Blind signatures

So, let's start by introducing hash functions and their main scopes. Then we will go deeper into categorizing digital signature algorithms.

A basic explanation of hash functions

Hash functions are widely used in cryptography for many applications. As we have already seen in *Chapter 3, Asymmetric Algorithms*, one of these applications is *blinding* an exchanged message when it is digitally signed. What does this mean? When Alice and Bob exchange a message using any asymmetric algorithm, in order to identify the sender, it commonly requires a signature. Generally, we can say that the signature is performed on the message [M]. For reasons we will learn later, it's discouraged to sign the secret message directly, so the sender has to first transform the message [M] into a function (M'), which everyone can see. This function is called the **hash of M**, and it will be represented in this chapter with these notations: f(H) or h[M].

We will focus more on the relations between hashes and digital signatures later on in this chapter. However, I have inserted hash functions in this chapter alongside digital signatures principally because hash functions are crucial for signing a message, even though we can find several applications related to hash functions outside of the scope of digital signatures, such as applications linked to the blockchain, like *distributed hash tables*. They also find utility in the search engine space.

So, the first question we have is: what is a hash function?

The answer can be found in the meaning of the word: *to reduce into pieces*, which in this case refers to the contents of a message or any other information being reduced into a smaller portion.

Given an arbitrary-length message [M] *as input, running a hash function* f(H), *we obtain an output (message digest) of a determined fixed dimension* (M').

If you remember the bit expansion *(Exp-function)*, seen in *Chapter 2, Introduction to Symmetric Encryption*, this is conceptually close to hashes. The bit expansion function works with a given input of bits that has to be expanded. We have the opposite task with hash functions, where the input dimension is bigger than the hash's bit value.

We call a hash function (or simply a hash) a **unidirectional function**. We will see later that this property of being *one-way* is essential in classifying hash functions.

Being a unidirectional or one-way function means that it is easy to calculate the result in one direction, but very difficult (if not impossible) to get back to the original message from the output of the function.

Let's see the properties verified in a hash function:

- Given an input message [M], its digest message h(M) can be calculated *very quickly*.

- It should be almost *impossible* (-%) to come back from the output (M') calculated through h(M) to the original message [M].

- It should be computationally *intractable* to find two different input messages [m1] and [m2], such that:

      h(m1) = h(m2)

In this case, we can say that the f(h) function *is strongly collision-free.*

To provide an example of a hash function, if we want the message digest of the entire content of Wikipedia, it becomes a fixed-length bit as follows:

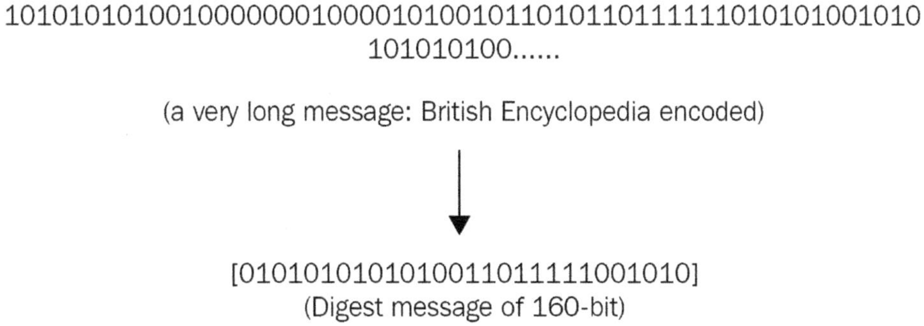

10101010100100000001000010100101101011011111010101001010
101010100......

(a very long message: British Encyclopedia encoded)

$$\downarrow$$

[010101010101001101111001010]
(Digest message of 160-bit)

Figure 4.1 – Example of hash digest

Another excellent metaphor for hash functions could be a funnel mincer like the following, which digests plaintext as input and outputs a fixed-length hash:

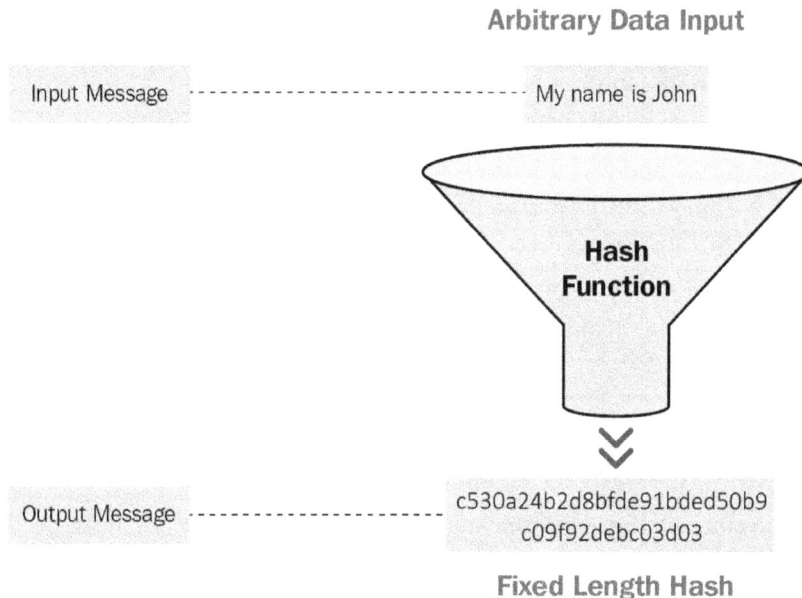

Arbitrary Data Input

Input Message ------------------------------------ My name is John

Hash Function

Output Message ------------------------- c530a24b2d8bfde91bded50b9
c09f92debc03d03

Fixed Length Hash

Figure 4.2 – Hash functions symbolized by a "funnel mincer"

The next question is: why and where are hash functions used?

Recalling what we said at the beginning, in cryptography, hash functions are widely used for a range of different scopes:

- Hash functions are commonly used in asymmetric encryption to avoid exposing the original message [M] when collecting digital signatures on it (we will see this later in this chapter). Instead, the hash of the original message proves the identity of the transmitter using (M') as a surrogate.

- Hashes are used to check the integrity of a message. In this case, based on the digital hash of the original message (M'), the receiver can easily detect whether someone has modified the original message [M]. Indeed, by changing only 1 bit among the entire content of the British Encyclopedia [M], for example, its hash function, h(M), will have a completely different hash value h(M') than the previous one. This particularity of hash functions is essential because we need a strong function to verify that the original content has not been modified.

- Furthermore, hash functions are used for indexing databases. We will see these functions throughout this book. We will use them a lot in the implementation of our *Crypto Search Engine* detailed in *Chapter 9, Crypto Search Engine*, and will also see them in *Chapter 8, Quantum Cryptography*. Indeed, hash functions are candidates for being *quantum resistant*, which means that hash functions can overcome a quantum computer's attack *under certain conditions*.

- The foundational concept of performing a secure hash function is that given an output h(M), it is difficult to get the original message from this output.

A function that respects such a characteristic is the *discrete logarithm,* which we have already seen in our examination of asymmetric encryption:

```
g^[a] ≡ y (mod p)
```

Where, as you may remember, given the output (y) and also knowing (g), it is very difficult to find the secret private key [a].

However, a function like this seems to be too slow to be considered for practical implementations. As you have seen above, one of the particularities of hashes is to be quick to calculate. As we have seen what hash functions are and their characteristics, now we will look at the main algorithms that implement hash functions.

Overview of the main hash algorithms

Since a hash algorithm is a particular kind of mathematical function that produces a fixed output of bits starting from a variable input, it should be collision-free, which means that it will be difficult to produce two hash functions for the same input value and vice versa.

There are many types of hash algorithms, but the most common and important ones are MD5, SHA-2, and CRC32, and in this chapter, we will focus on the **Secure Hash Algorithm (SHA)** family.

Finally, for your knowledge, there is **RIPEMD-160**, an algorithm developed by EU scientists in the early 1990s available in 160 bits and in other versions of 256 bits and 320 bits. This algorithm didn't have the same success as the SHA family, but could be the right candidate for security as it has never been broken so far.

The SHA family, developed by the **National Security Agency (NSA)**, is the object of study in this chapter. I will provide the necessary knowledge to understand and learn how the SHA family works. In particular, **SHA-256** is currently the hash function used in **Bitcoin** as **Proof of Work (PoW)** when mining cryptocurrency. The process of creating new Bitcoins is called mining; it is done by solving an extremely complicated math problem, and this problem is based on SHA-256. At a high level, Bitcoin mining is a system in which all the transactions are devolved to the miners. Selecting one megabyte worth of transactions, miners bundle them as an input into SHA-256 and attempt to find a specific output the network accepts. The first miner to find this output receives a reward represented by a certain amount of Bitcoin. We will look at the SHA family in more detail later in this book, but now let's go on to analyze at a high level other hash functions, such as MD5.

We can start by familiarizing ourselves with hash functions, experimenting with a simple example of an MD5 hash.

You can try to use the **MD5 Generator** to hash your files. My name converted into hexadecimal notation with MD5 is as follows:

Massimo Bertaccini
(Hash MD5)

=

f38e1056801af5d079f95c48fbfd2d60
(Bit conversion)
1111001110001110000100000101011010000000000110101111 0101
1101000001110011111100101011100010010001111101 1111111010
010110101100000

Figure 4.3 – MD5 hash function example

As I told you before, the MD family was found to be insecure. Attacks against MD5 have been demonstrated, valid also for RIPEMD, based on differential analysis and the ability to create collisions between two different input messages.

So, let's go on to explore the mathematics behind hash functions and how they are implemented.

Logic and notations to implement hash functions

This section will give you some more knowledge about the operations performed inside hash functions and their notations.

Recalling Boolean logic, I will outline more symbols here than those already seen in *Chapter 2, Introduction to Symmetric Encryption*, and shore up some basic concepts.

Operating in Boolean logic means, as we saw in *Chapter 2, Introduction to Symmetric Encryption*, performing operations on bits. For instance, taking two numbers in decimal notation and then transposing the mathematical and logical operations on their corresponding binary notations, we get the following results:

- X ∧ Y = AND **logical conjunction** – this is a bitwise multiplication (mod 2). So, the result is 1 when both the variables are 1, and the result is 0 in all other cases:

```
X=60   ---------->     00111100
Y=240  ---------->     11110000
AND= 48--------->      00110000
```

- X ∨ Y = OR **logical disjunction** – a bitwise operation (mod 2) in which the result is 1 when we have at least 1 as a variable in the operation, and the result is 0 in other cases:

```
X=60   ---------->     00111100
Y=240  ---------->     11110000
OR = 252 ------>       11111100
```

- X ⊕ Y = XOR **bitwise sum (mod 2)** – we have already seen the XOR operation. XOR means that the result is 1 when bits are different, and the result is 0 in all other cases:

```
a=60   ---------->     00111100
b=240  --------->      11110000
XOR=204 ------->       11001100
```

Other operations useful for implementing hash functions are the following:

- ¬X: This operation (the NOT or *inversion* operator ~) converts the bit 1 to 0 and 0 to 1.

    So, for example, let's take the following binary string:

    ```
    01010101
    ```

It will become the following:

```
10101010
```

- X << r: This is the shift left bit operation. This operation means to shift (X) bits to the left of r positions.

In the following example, you can see what happens when we shift to the left by 1 bit:

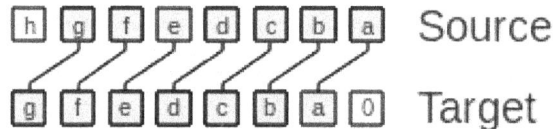

Figure 4.4 – Scheme of the shift left bit operation

So, for example, we have decimal number 23, which is 00010111 in byte notation. If we do a shift left, we get the following:

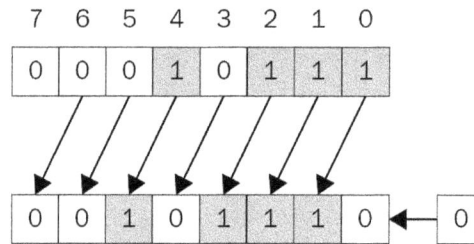

Figure 4.5 – Example of the shift left bit operation (1 position)

The result of the operation after the shift left bit is (00101110) 2 = 46 in decimal notation.

- X >> r: This is the shift right bit operation. This is a similar operation as the previous, but it shifts the bits to the right.

The following diagram shows the scheme of the shift right bit operation:

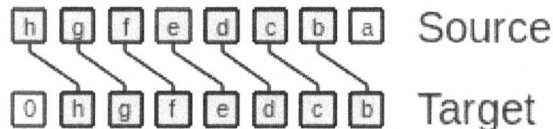

Figure 4.6 – Scheme of the shift right bit operation (1 position)

As before, let's consider this for the decimal number 23, which is 00010111 in byte notation. If we do a shift right bit operation, we will get the decimal number 11 as the result, as you can see in the following example:

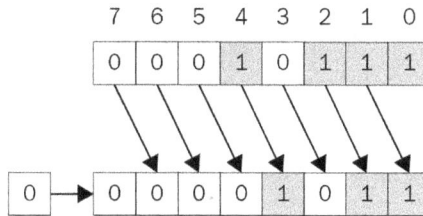

Figure 4.7 – Example of the shift right bit operation (1 position)

- X↺r: This is left bit rotation. This operator (also represented as <<<) means a circular rotation of bits, similar to shift, but with the key difference here that the initial part becomes the final part. It's used in the SHA-1 algorithm to rotate the variables A and B, respectively, by 5 and 30 positions (A<<<5; B<<<30) as will be further explained in the following section.

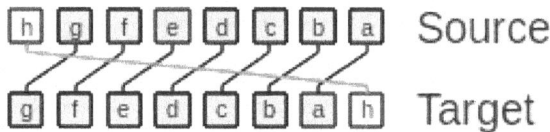

Figure 4.8 – Scheme of left bit rotation (1 position)

If we apply left bit rotation to our example of the number 23 as expressed in binary notation, we get the following:

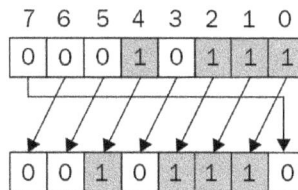

Figure 4.9 – Example of left bit rotation

In this case, the result of the left bit rotation operation is the same as the shift left bit operation: $(00101110)_2 = 46$. But, if we perform a right bit rotation, we will get the following:

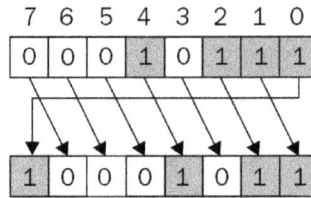

Figure 4.10 – Example of right bit rotation

In this case, the result of the operation is $(10001011)_2 = 139$ expressed in decimal notation.

- ⊞: This operator represents the modular sum $(X+Y)$ $(mod\ 2^32)$ and is used in SHA to represent this operation, as you will see later in our examination of the SHA-1 algorithm (*Figure 4.14*).

After having analyzed the logic operators used to perform hash functions, two more issues have to be considered in order to implement hash functions:

- In SHA, for example, there are some constants (Kt) that go from 0 to n (we will look at this in the following section).

- The notations are generally expressed in hexadecimal.

Let's now see how the hexadecimal system works.

It uses the binary numbers for 0 to 9, consisting of 4 bits per number, for example:

```
0= 0000
1= 0001
2= 0010
.........
9= 1001
```

Then, from 10 to 16 we have 6 capital letters, A, B, C, D, E, F, to reach a total of 16 (hex) numbers:

```
A= 1010
B= 1011
......... .
F= 1111
```

For a better and clearer understanding, in the following figure, you can see a comparison between the binary, hexadecimal, and decimal systems:

Binary	Hexadecimal	Decimal
$b=2$	$b=16$	$b=10$
0000	0	0
0001	1	1
0010	2	2
0011	3	3
0100	4	4
0101	5	5
0110	6	6
0111	7	7
1000	8	8
1001	9	9
1010	A	10
1011	B	11
1100	C	12
1101	D	13
1110	E	14
1111	F	15

Figure 4.11 – Comparison between binary, hexadecimal, and decimal systems

The hexadecimal system is used to represent bytes, where 1 byte is an 8-digit binary number.

For example, 1 byte (8 bits) is expressed as $(1111\ 0010)2 = [F2]16 = 242$.

Now that we have the instruments to define a hash function, let's go ahead and implement SHA-1, which is the simplest model in the SHA family.

Explanation of the SHA-1 algorithm

Secure Hash Algorithm 1 (SHA-1) was designed by the NSA. Since 2005, it has been considered insecure and was replaced by its successors, SHA-2 and SHA-3. In this section, we will examine SHA-1 simply as a case study to better understand how hash functions are implemented.

SHA-1 returns a 160-bit output based on an iterative procedure. This concept will become clearer in the next few lines.

As for other hash functions, the message [m] made of variable bit input, is broken into 512-bit fixed-length blocks: m= [m1,m2,m3,....ml].

In the last part of this section, you will see how an input message [m] of 2,800 bits will be transformed into blocks [m1,m2,.....ml] of 512 bits each.

The blocks are elaborated through a compression function f (H) (it will be better analyzed later in *Step 3* of the algorithm) that combines the current block with the result obtained in the previous round. There are four rounds and they correspond to the variable (t), whose range is divided into four t-rounds as shown in *Figure 4.12*, each one made of 20 steps (for a total of 80 steps). Each iteration can be seen as a counter that runs along the values of each range made of 20 values. As you can see from *Figures 4.12* and *4.13*, each iteration uses the constants (Kt) and the operations ft (B,C,D) of the corresponding round.

Each round updates the sub-registers (A,B,C,D,E) after the other. At the end of the 4th round, when t = 79, the sub-registers (A,B,C,D,E) are added to the sub-registers (H0, H1, H2, H3, H4) to perform the 160 bits final hash value.

Let's now examine the issue of constants in SHA-1.

Constants are fixed numbers expressed in hexadecimal defined with particular criteria. An important criterion adopted to choose the constants in SHA is to avoid collisions. Collisions happen when a hash function gives the same result for two different blocks starting from different constants. So, even though someone might think it would be a good idea to change the values of the constants, don't try to change them because that could cause a collision problem.

In SHA-1, for example, the given constants are as follows:

$$Kt = \begin{cases} \text{5A827999} & \text{if} & 0 \leq t \geq 19 \\ \text{6ED9EBA1} & \text{if} & 20 \leq t \geq 39 \\ \text{8F1BBCDC} & \text{if} & 40 \leq t \geq 59 \\ \text{CA62C1D6} & \text{if} & 60 \leq t \geq 79 \end{cases}$$

Figure 4.12 – The Kt constants in SHA-1

Where, in different ranges of (t), different values correspond with the constants (Kt), as you see in the preceding figure.

Besides the constants, we have the function ft (B,C,D) defined as follows:

$$
f_t \ (B, C, D) = \begin{cases} (B \wedge C) \vee ((\neg B) \wedge D) & \text{if} \quad 0 \le t \ge 19 \\ B \oplus C \oplus D & \text{if} \quad 20 \le t \ge 39 \\ (B \wedge C) \vee (B \wedge D) \vee (C \wedge D) & \text{if} \quad 40 \le t \ge 59 \\ B \oplus C \oplus D & \text{if} \quad 60 \le t \ge 79 \end{cases}
$$

Figure 4.13 – The ft function

The first initial register of SHA-1 is X0, a 160-bit hash function generated by five sub-registers (H0, H1, H2, H3, H4) consisting of 32 bits each. We will see the initialization of these sub-registers in *Step 2* using constants expressed in hexadecimal numbers.

Now let's explain the SHA-1 algorithm by dividing the process into four steps to obtain the final hash value of 160 bits:

- **Step 1** – Starting with a message [m], operate a concatenation of bits such that:

  y= m1 || m2 || m3 || … ||mL

  Where the || symbol stands for the concatenation of bits expressed by each block of message [ml] consisting of 512 bits.

- **Step 2** – Initialization of the sub registers: H0 = 67452301, H1 = EFCDAB89, H2 = 98BADCFE, H3 = 10325476, H4 = C3D2E1F0.

  You may notice that these constants are expressed in hexadecimal notation. They were chosen by the NSA, the designers of this algorithm.

- **Step 3** – For j = 0, 1, … , L-1, execute the following instructions:

  a) mi = W0 || W1………. . ||W15,

  where each (Wj) consists of 32 bits.

  b) For t = 16 to 79, put:

  Wt = (Wt-3 ⊕ Wt-8 ⊕ Wt-14 ⊕ Wt-16) ⊋ 1

  c) At the beginning, we put:

  A = H0, B = H1, C = H2, D = H3, E = H4

Each variable (A,B,C,D,E) is of 32 bits length for a total length of the sub-register of 160 bits.

d) For 80 iterations, where $0 \leq t \leq 79$, execute the following steps in succession:

```
T = (A ↺ 5) + ft((B,C, D) +E + Wt + KtE = D, D = C, C =
(B ↺ 30), B = A, A = T
```

e) The sub-registers (A,B,C,D,E) are added to the sub-registers (H0, H1, H2, H3, H4):

```
H0 = H0 + A, H1 = H1 + B, H2 = H2 + C, H3 = H3 + D, H4 =
H4 + E
```

- **Step 4** – Take the following as output:

```
H0 || H1 || H2 || H3 || H4.
```

This is the hash value of 160 bits.

Here you can see a scheme of SHA-1 (the sub-registers are A, B, C, D, E):

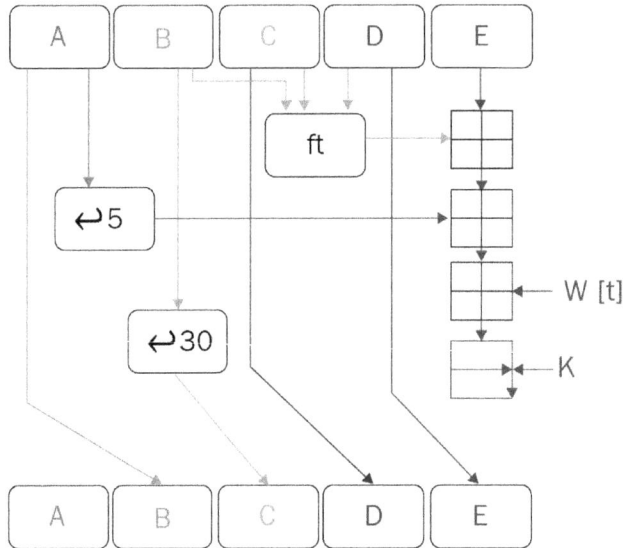

Figure 4.14 – SHA-1 scheme of the operations in each sub-register

> **Important Note**
>
> Remember from the previous section on the logic and notations that:
>
> X ↻ r is left bit rotation, also represented as <<<, and means a circular rotation of bits. So, in the case of A ↻ 5, it means that bits rotate 5 positions to the left, while in the case of B ↻ 30 bits, bits rotate 30 positions to the left.
>
> ⊞: This operator represents the modular sum (X+Y) (mod 2^32).

Notes and example on SHA-1

In this section, we will analyze the SHA-1 algorithm in a little more detail.

Since *Steps 1* and *2* are just message initialization and sub-register initialization, the algorithm's core is in *Step 3*, where you can see a series of mathematical operations consisting of bit concatenation, XOR, bit shift, bit transposition, and bit addition.

Finally, *Step 4* reduces the output to 160 bits just because each sub-register H0, H1, H2, H3, and H4 is 32 bits. In *Step 1*, we mentioned that the minimum block message provided in the input has to be 512 bits. The message [m] could be divided into blocks of 512 bits, and if the original message is shorter than 512 bits, we have to apply a padding operation, involving the addition of bits to complete the block.

SHA-1 is used to compute a message digest of 160-bit length for an arbitrary message that is provided as input. The input message is a bit string, so the length of the message is the number of bits that make up the message (that is, an empty message has length 0). If the number of bits in a message is a multiple of 8 (a byte), for compactness, we can represent the message in hexadecimal. Conversely, if the message is not a multiple of a byte, then we have to apply padding. The purpose of the padding is to make the total length of a padded message a multiple of 512. As SHA-1 sequentially processes blocks of 512 bits when computing the message digest, the trick to getting a strong message digest is that the hash function has to provide the best grade of confusion/diffusion on the bits involved in the iterative operations.

The process of padding consists of the following sequence of passages:

1. SHA-1 starts by taking the original message and appends 1 bit followed by a sequence of 0 bits.

2. The 0 bits are added to ensure that the length of the new message is a multiple of 512 bits.

3. For this scope, the new message will have a final length of n*512.

For example, if the original message has a length of 2800 bits, it will be padded with 1 bit at the end followed by 207 bits. So, we will have 2800 + 1 + 207 = 3008 bits. Then to make the result of the padding a multiple of 512, using the division algorithm, notice that 3008 = 5 * 512 + 448, so to get to a multiple of 512, we pad the message with 64 zeros. So, we finally obtain 3008 + 64 = 3072, which is the number of bits of the padded message, divisible by 512 (bits per block).

Example of one block encoded with SHA- 1

Let's understand this with the help of a practical example:

- **Step 1 – Padding the message**:

    Suppose we want to encode the message abc using SHA-1, which in binary system is expressed as:

    ```
    abc = 01100001 01100010 01100011
    ```

    In hex, the string is:

    ```
    abc = 616263
    ```

    As you can see in the following figure, the message is padded by appending 1, followed by enough 0s until the length of the message becomes 448 bits. Since the message abc is 24 bits in length, 423 further bits are added. The length of the message represented by 64 bits is then added to the end, producing a message that is 512 bits long:

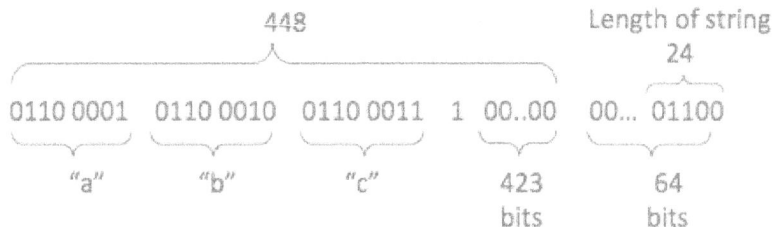

Figure 4.15 – Padding of the message in SHA-1

- **Step 2 – Initialization of the sub-registers**:

  The initial hash value for the sub-registers H0, H1, H2, H3, and H4 will be:

  ```
  H[0] = 67452301
  H[1] = EFCDAB89
  H[2] = 98BADCFE
  H[3] = 10325476
  H[4] = C3D2E1F0
  ```

- **Step 3 – Block contents**:

  ```
  W[0] = 61626380 W[1] = 00000000 W[2] = 00000000 W[3] =
  00000000 W[4] = 00000000 W[5] = 00000000 W[6] = 00000000
  W[7] = 00000000 W[8] = 00000000 W[9] = 00000000 W[10]
  = 00000000 W[11] = 00000000 W[12] = 00000000 W[13] =
  00000000 W[14] = 00000000 W[15] = 00000018
  ```

  Iterations on the sub-registers:

  ```
                 A        B        C        D        E
  t = 0: 0116FC33 67452301 7BF36AE2 98BADCFE 10325476
  t = 1: 8990536D 0116FC33 59D148C0 7BF36AE2 98BADCFE
  . . . . . . . . .
  Tt = 79: 42541B35 5738D5E1 21834873 681E6DF6 D8FDF6AD
  ```

  Addition of the sub-registers:

  ```
  H[0] = 67452301 + 42541B35 = A9993E36
  H[1] = EFCDAB89 + 5738D5E1 = 4706816A
  H[2] = 98BADCFE + 21834873 = BA3E2571
  H[3] = 10325476 + 681E6DF6 = 7850C26C
  H[4] = C3D2E1F0 + D8FDF6AD = 9CD0D89D
  ```

- **Step 4 – Result**:

    After performing the four rounds, the final message digest of the string abc of 160-bit hash is:

    ```
    A9993E36 4706816A BA3E2571 7850C26C 9CD0D89D
    ```

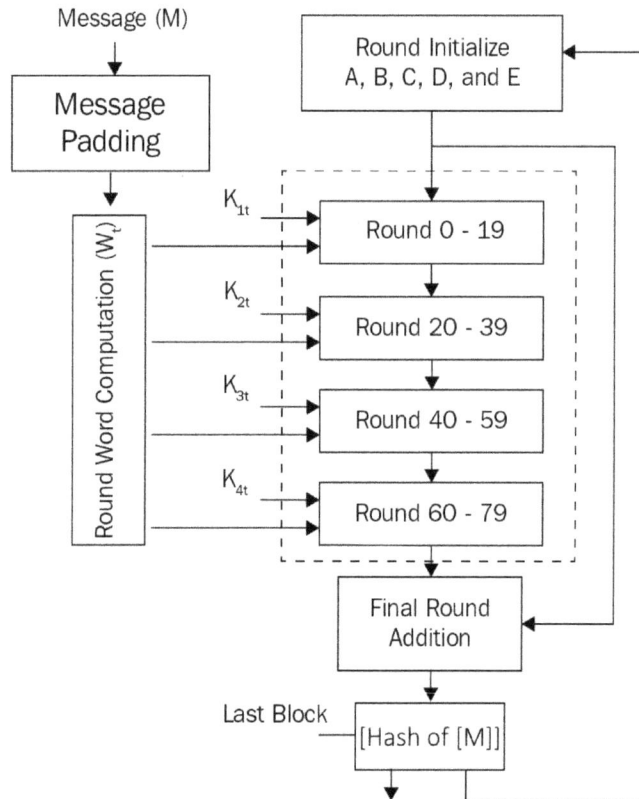

Figure 4.16 – A complete round in SHA-1

Now that we have learned about hash functions, we are prepared to explore digital signatures.

Authentication and digital signatures

In cryptography, the **authentication** problem is one of the most interesting and difficult problems to solve. Authentication is one of the most sensitive functions (as well as the most used) for the procedure of access control.

Authentication is based on three methods:

- On something that *only the user knows* (for example, a password)

- On something that *only the user holds* (a smart card, device, or a token)

- On something that *characterizes the user* (for example, fingerprints, an iris scan, and in general biometric characteristics of a person)

In addition to these three methods, there is one more method involving something that *only the user holds, related to something that uniquely characterizes them*: brain waves. For example, when you look at a picture or think about it, the brain waves activated by your brain are unique, different from any other brain waves that characterize another person.

This invention is related to the authentication between human/computer. In other words, with Brain Password, it is possible to unlock devices such as computers and telephones and get access to web applications. I will introduce brain passwords in *Chapter 6, New Algorithms in Public/Private Key Cryptography*.

In the next section, we will analyze an authentication method based on digital signatures. Keep in mind (as we will see later on) that there are similar methods of authentication based, for example, on zero knowledge, which will be covered further in *Chapter 5, Introduction to Zero-Knowledge Protocols*.

Let's see how a method of authentication based on a digital signature over public/private key encryption works.

Let's consider an example where Alice wants to transmit a message to Bob. She knows Bob's public key (as in RSA), so Alice encrypts the message [M] with RSA and sends it to Bob.

Let's look at some of the common problems faced when Alice transmits the message:

1. How can Bob be sure that this message comes from Alice (*authentication problem*)?

2. After Bob has received the message, Alice could always deny that she is the transmitter (non-*repudiation*).

3. Another possible issue is that the message [M] could be manipulated by someone who intercepted it (a **Man-in-the-Middle** (**MiM**) attack) and subsequently, the attacker could change part of the content. This is a case of *integrity loss* of the message.

4. Finally, here's another consideration (and probably the first time you'll see it in a textbook): the message could be intercepted and *spied* on by the attacker, who recovers the content but decides not to modify it. How can the receiver and the transmitter be sure that this doesn't happen? For example, how can you be sure that your telecommunication provider or the cloud that hosts your data don't spy on your messages? I refer to this as the spying problem.

To answer the last question, you could rightly say that since the message [M] is encrypted, it is difficult – if not impossible – for the attacker to recover the secret message without knowing the key.

I will demonstrate that (under some conditions) it is possible to spy on an encrypted message, even if the attacker doesn't know the secret key for decryption.

We will discuss problems *1*, *2*, and *3* in this chapter. As for problem *4*, I will explain an attack method for spying and the relative way to repair this problem in *Chapter 6, New Algorithms in Public/Private Key Cryptography*.

A digital signature is a proof that the sender of the message [M] instructs the receiver to do the following:

- Prove their identity
- Prove that the message [M] has not been manipulated
- Provide for the non-repudiation of the message

Let's see how digital signatures can help us to avoid all these problems and also how it is possible to sign a message based on the type of algorithm and the different ways of using signatures.

We will first look at RSA digital signatures, and will then analyze the other methods of signatures in public-private key algorithms.

RSA digital signatures

As I have told you before, it's possible to sign a message in different ways; we will now explore how to mathematically sign one and how anyone can verify a signature.

Recalling the RSA algorithm from *Chapter 3, Asymmetric Encryption*, we know that the process of encryption for Bob is as follows:

Perform encryption based on the public key of Alice (Na) and define the secret elements within the square brackets [M]:

```
[M]^e ≡ c (mod Na)
```

(Na) is the public key of the receiver (Alice).

Therefore, Bob will encrypt the message [M] with Alice's public key, and Alice will decrypt [M] with her private key [da] by performing the following operation:

```
c^[da] ≡ [M] (mod Na)
```

Where the parameter [da] is the private key of Alice, given by the operation:

```
INV (e) ≡ [da] mod(p-1)(q-1)
```

The process for the digital signing and verification of Bob's identity is as follows:

- **Step 1**: Bob chooses two big prime numbers, [p1, q1], and keeps them secret.
- **Step 2**: Bob calculates Nb = [p1] * [q1] and publishes (Nb) as his public key.
- **Step 3**: Bob calculates his private key [db] such that:

  ```
  INV (e) ≡ [db] mod (p1-1)(q1-1)
  ```

- **Step 4**: Bob performs the signature:

  ```
  S ≡ [M]^[db] (mod Nb)
  ```

  Where (S) and (Nb) are public and [db] is secret.

  What about [M]?

  [M], the message, is supposed to be secret and shared only between Bob and Alice. However, a digital signature should be verified by everyone. We will deal with this issue later on. Now let's verify the signature (S).

- **Step 5**: Verifying the signature:

  Signature verification is the inverse process of the above. Alice (or anyone) can verify the following:

  ```
  S^e ≡ [M] (mod Nb)
  ```

If the preceding equation is verified, then Alice accepts the message [M].

Before analyzing the issue of [M], let's understand this algorithm with the help of a numerical example.

Numerical example:

Recalling the RSA example from *Chapter 3, Asymmetric Encryption*, we have the following numerical parameters given by the RSA algorithm:

```
[M] = 88
e = 9007
```

Alice's parameters were as follows:

```
[p] = 101
[q] = 67
Na = 6767
```

Let's perform all of the steps of RSA to show the comprehensive process of digitally signing the message:

- **Step 1** – Bob's encryption is as follows:

```
[M] ^e ≡ c (mod N)
88^9007 ≡ 6621 (mod 6767)
c = 6621
```

- **Step 2** – Bob's signature (S) for the message [M] will be generated as follows:

Bob chooses two prime numbers, [p, q]:

```
[p] = 211
[q] = 113
Nb= 211*113 = 23843
9007 * [db] ≡ 1 (mod (211-1)*(113-1))
```

Solving the modular equation for [db], we have the following:

```
[db] = 9103
```

Now, Bob can sign the message:

```
[M]^[db] ≡ S (mod Nb)
88^9103 ≡ 19354 (mod 23843)
S = 19354
```

Bob sends to Alice the pair (c, S) = (6621, 19354).

- **Step 3** – Alice's decryption will be as follows:

```
c^[da] ≡ [M] (mod Na)
```

Alice calculates [da]:

```
9007 * [da] ≡ 1 (mod (101-1)*(67-1))
[da] = 3943
```

Alice decrypts the cryptogram (c) and obtains the message [M]:

```
c^[da] ≡ [M] (mod Na)
6621^3943 ≡ 88 (mod 6767)
```

- **Step 4** – Verification of Bob's identity:

If the signature (S) elevated to the parameter (e) gives the message [M], then Alice can be sure that the message was sent by Bob:

```
S^e ≡ [M] (mod Nb)
19354^9007 = 88 (mod 23843)
```

Indeed, Alice obtains the message M=88.

Important Note

I hope someone has noticed that the message [M] can be verified by anyone, and not only by Alice, because (S, e, Nb) are public parameters. So, if Bob uses the original message [M], instead of its hash h[M] to gain the signature (S) (see above for the encryption procedure), everyone who knows Bob's public key could easily recover the message [M]!

It's just a matter of solving the following equation to recover the secret message [M]:

```
S^e ≡ x (mod Nb)
```

Where the parameters (S, e, Nb) are all known.

In this case, *hash functions* come to our aid. Performing the hash function h[M] = (m), Bob will send the couple (c, S), signing (m) instead of [M].

Alice already got the encrypted message [M], so only Alice can verify it:

```
h[M] = m
```

If it is TRUE, then Bob's identity will be verified by Alice; if it is FALSE, Bob's claim of identity will be refused.

Why do digital signatures work?

If anyone else who isn't Bob tries to use this signature, they will struggle with the *discrete logarithm* problem. Indeed, generating [db] in the following equation is a hard problem for the attacker to solve:

```
m^[db] ≡ S (mod Nb)
```

Also, even if (m, S) are known by the attacker, it is still very difficult to calculate [db]. In this case, we are dealing with a discrete logarithm problem like what we saw in *Chapter 3, Asymmetric Algorithms*.

Let's try to understand what happens if Eve (an attacker) tries to modify the signature with the help of an example:

Eve (the attacker) exchanges [db] with (de), computing a fake digital signature (S'):

```
m^(de) ≡ S' (mod Nb)
```

If Eve is able to trick Alice to accept the signature (S'), then Eve can make an MiM attack by pretending to be Bob, substituting (S') with the real signature (S).

But when Alice verifies the signature (S'), she recognizes that it doesn't correspond to the correct signature performed by Bob because the hash of the message is (m'), not (m):

```
(S')^e ≡ m' (mod Nb)
m' ≢ m
```

So, Alice refuses the digital signature.

That is why cryptographers also need to pay a lot of attention to the collisions between hashes.

Now that we have got a different result, that is, (m') instead of (m), Alice understands there is a problem and doesn't accept the message [M].

This is the scope of the signature (S).

The preceding attack is a simple trick, but there are some more intelligent and sneaky attacks that we will see later on, in *Chapter 6, New Algorithms in Public/Private Key Cryptography*, when I will explain unconventional attacks.

Digital signatures with the ElGamal algorithm

ElGamal is a public-private key algorithm, as we saw in *Chapter 3, Asymmetric Algorithms*, based on the *Diffie-Hellman key exchange.*

In ElGamal, we have different ways of signing the message than RSA, but all are equally valid.

Recalling the ElGamal encryption technique, we have the following elements:

- (g, p) : Public parameters
- $[k]$: Alice's private key
- $[M]$: The secret message
- $B \equiv g^{\wedge}[b] \pmod{p}$: Bob's public key
- $A \equiv g^{\wedge}[a] \pmod{p}$: Alice's public key

Alice's encryption:

```
y1 ≡ g^[k] (mod p)
y2 ≡ [M]*B^[k] (mod p)
```

> **Note**
> The elements inside the square brackets indicate secret parameters; all the others are public.

Now, if Alice wants to add her digital signature to the message, she will make a *hash of the message*, h[M], to protect the message [M], and will then transmit the result to Bob in order to prove her identity.

To sign the message [M], Alice first has to generate the hash of the message h[M]:

```
h[M] = m
```

Now, Alice can operate with the digest value of [M] —> (m) in cleartext because, as we have learned before, it is almost impossible to return to [M] from its cryptographic hash (m).

Alice calculates the signature (S) such that:

- **Step 1** – Making the inverse of [k] in (mod p-1):

    [INVk] ≡ k^(-1) (mod p-1)

- **Step 2** – Performing the equation:

    S ≡ [INVk]* (m - [a] *y1) (mod p-1)

    Alice sends to Bob the public parameters (m, y1, S).

- **Step 3** – In the first verification, Bob performs V1:

    V1 ≡ A^(y1) * y1^(S) (mod p)

    After the decryption step, if h[M] = m, Bob obtains the second parameter of verification, V2:

    V2 ≡ g^m (mod p)

    Finally, if V1 = V2, Bob accepts the message.

Now, let's better understand this algorithm with the help of a numerical example.

Numerical example:

Let's suppose that the value of the secret message is:

    [M] = 88

We assign the following values to the other public parameters:

```
g = 7
p = 200003
h[M] = 77
```

The first step is the "Key initialization" of the private and public keys:

```
[b] = 2367 (private key of Bob)
[a] = 5433 (private key of Alice)
[k] = 23 (random secret number of Alice)
B = 151854 (public key of Bob)
A = 43725 (public key of Alice)
y1 ≡ g^[k] (mod p) = 7^23 (mod 200003) = 90914
```

After the initialization process, Alice calculates the inverse of the key (*Step 1*) and then the signature (*Step 2*):

- **Step 1** – Alice computes the inverse of [k] in (mod p-1):

```
[INVk] ≡ [k]^(-1) (mod p-1) = 23^-1 (mod 200003 - 1) =
34783
```

- **Step 2** – Alice can get now the signature (S):

```
S ≡ [INVk]* (m - [a] *y1) (mod p-1)
S ≡ 34783 * (77 - 5433 * 90914) (mod 200003 - 1)
S = 72577
```

Alice sends to Bob the public parameters (m, y1, S) = (77, 90914, 72577)

- **Step 3** – With those parameters Bob can perform the first verification (V1). Consequently, he computes V2. If V2 = V1 Bob accepts the digital signature (S):

```
V1 ≡ A^(y1) * y1^(S) (mod p)
V1 ≡ 43725 ^ (90914) * 90914 ^ 72577 (mod 200003) = 76561
V1 = 76561
```

Bob's verification (V2):

```
V2 ≡ g^m (mod p)
V2 ≡ 7^ 77 (mod 200003) = 76561
V2 = 76561
```

Bob verifies that V1= V2.

Considering the underlying problem that makes this algorithm work, we can say that it is the same as the discrete logarithm. Indeed, let's analyze the verification function:

```
V1 ≡ A^(y1) * y1^(S) (mod p)
```

All the elements are made by discrete powers, and as we already know, it's a hard problem (for now) to get back from a discrete power even if the exponent or the base is known. It is not sufficient to say that *discrete powers and logarithms* ensure the security of this algorithm. As we saw in *Chapter 3, Asymmetric Cryptography*, the following function could also be an issue:

```
y2 ≡ [M]*B^[k] (mod p)
```

It's given by multiplication. If we are able to recover $[k]$, then we have discovered $[M]$.

So, the algorithm suffers not only from the discrete logarithm problem but also from the factorization problem.

Now that you have learned about the uses and implementations of digital signatures, let's move forward to explore another interesting cryptographic protocol: blind signatures.

Blind signatures

David Chaum invented **blind signatures**. He struggled a lot to find a cryptographic system to anonymize *digital payments*. In 1990, David funded *eCash*, a system that adopted an untraceable currency. Unfortunately, the project went bankrupt in 1998, but Chaum will be forever remembered as one of the pioneers of digital money and one of the fathers of modern cryptocurrency, along with Bitcoin.

The underlying problems that Chaum wanted to solve were the following:

- To find an algorithm that was able to avoid the *double-spending problem* for electronic payments.

- To make the digital system *secure and anonymous* to guarantee the *privacy* of the user.

In 1982, Chaum wrote an article entitled *Blind Signatures for Untraceable Payments*. The following is an explanation of how the blind signatures described in the article work and how to implement them.

Signing a message *blind* means to sign something without knowing the content. It could be used not only for digital payments but also if Bob, for example, wants to publicly register something that he created without making known to others the details of his invention. Another application of blind signatures is in electronic voting machines, where someone makes a choice (say, in an election for the president or for a party, for example). In this case, the result of the vote (the transmitted message) has to be known by the receiver obviously, but the identity of the voter has to remain a secret if the voter wants to be sure that their vote will be counted (that is the proper function of blind signatures).

I will expose an innovative blind signature scheme for the *MBXI cipher* in *Chapter 6, New Algorithms in Public/Private Key Cryptography*, where I will introduce new ciphers in private/public keys, including the MBXI, invented and patented by me in 2011.

Let's see now how David Chaum's protocol works by performing a blind signature with RSA.

Blind signature with RSA

Suppose Bob has an important secret he doesn't want to expose to the public until a determined date. For example, he discovered a formidable cure for coronavirus and he aspires to get the Nobel Prize.

Alice represents the commission for the Nobel Prize.

Alice picks up two big secret primes [pa, qa]:

- [pa*qa] = Na, which is Alice's public key.
- (e) is the same public parameter already defined in RSA.

The parameter [da] is Alice's private key, given by this operation:

```
INV (e) ≡ [da] mod(pa-1)(qa-1)
```

Suppose that [M1] is the secret belonging to Bob. I have just called it [M1] to distinguish it from the regular [M].

Bob picks up a random number [k] and keeps it secret.

Now Bob can go ahead with the blind signature protocol on [M1]:

- **Step 1** – Bob performs encryption (t) on [M1] to *blind* the message:

  ```
  t ≡ [M1] * [k]^e (mod Na)
  ```

  Bob sends (t) to Alice.

- **Step 2** – Alice can perform the blind signature given by the following operation:

  ```
  S ≡ t^[da] (mod Na)
  ```

  Alice sends (S) to Bob, who can verify whether the blind signature corresponds to the message [M1].

- **Step 3** – Verification:

  Bob calculates (V):
  ```
  S/k ≡ V (mod Na)
  ```

  Then he can verify the following:
  ```
  V^e ≡ [M1] (mod Na)
  ```

If the last operation is TRUE, it means Alice has effectively signed *blind* [M1]. In this case, Bob can be sure of the following:

```
[M1]^[da] ≡ V (mod Na)
```

Since no one except Alice could have performed function (S) without being able to solve the *discrete logarithm* problem (as already seen in *Chapter 3*, *Asymmetric Algorithms*), the signer must be Alice, for sure. This sentence remains valid until any other variable occurs; for example, when someone finds a logical way to solve the discrete logarithm or a quantum computer reaches enough qubits to break the algorithm, as we'll see in *Chapter 8*, *Quantum Cryptography*.

Let's see an example to better understand the protocol.

Numerical example:

1. The parameters defined by Alice are as follows:

    ```
    pa = 67
    qa = 101
    ```

    So, the public key (Na) is the following:

    ```
    67*101 = Na = 6767
    da ≡ 1/e (mod (pa-1) * (qa-1))
    Reduce [e*x == 1, x, Modulus -> (pa - 1)*(qa - 1)]
    [da] = 1553
    e = 17
    ```

2. Bob picks up a random number, [k]:

    ```
    k = 29
    ```

    Bob calculates (t):

    ```
    t ≡ M1* k^e ≡ 88 * 29^17 = 3524 (mod 6767)
    ```

3. Bob sends (t) to Alice ————————————> Alice can now blind-sign (t):

    ```
    S ≡ t^[da] ≡ 3524^1553 = 1533 (mod 6767)
    ```

4. Bob can verify (S) <————————— Alice sends (S = 1553) to Bob:

    ```
    S/k ≡ V (mod Na)
    1533/29 = 2853 (mod 6767)
    Reduce [k*x == S, x, Modulus -> Na]
    ```

```
x = 2853
V = 2853
if:
V^e ≡ [M1]  (mod Na)
2853^17 ≡ 88  (mod 6767)
```

Then, Bob accepts the signature (S).

As you can see from this example and the explanation of blind signatures, Alice is sure that Bob's discovery (the coronavirus cure) belongs to him, and Bob can preserve his invention without declaring the exact content of it before a certain date.

Notes on the blind signature protocol

You can do a double-check on [M1], so you will be able to realize that Alice has really signed [M1] without knowing anything about its value:

```
M1^[da] ≡ V  (mod Na)
88^1553 ≡ 2853  (mod 6767)
```

A warning about blind signatures is necessary, as Alice doesn't know what she is going to sign because [M1] is hidden inside (t). So, Bob could also attempt to convince Alice to sign a $1 million check. There is a lot of danger in adopting such protocols.

Another consideration concerns possible attacks.

As you see here, we are faced with a factorization problem:

```
t ≡ M1* [k]^e  (mod Na)
```

We can see that (t) is the product of [X*Y]:

```
X = M1 <———— Factorization problem
Y = k^e
```

It doesn't matter if the attacker is unable to determine [k], as they can always attempt to find [M1] by factoring (t), if [M1] is a small number, for example. It's simply the case of M1=0, because of course (t) will be zero and the message can be discovered by the attacker to be M = 0.

On the other hand, if we assume, for example, that (k^e) is a small number, since [k] is random, then the attacker can perform this operation:

```
Reduce [(k^e)*x == t, x, Modulus -> Na]
x = MESSAGE
```

In this case, if, unfortunately, $[k^e]$ (mod Na) results in a small number, the attacker can recover the message [M1].

As you will understand after reading the next chapter, blind signatures are the precursor to zero-knowledge protocols; the object of study in the next *Chapter 5, Introduction to Zero-Knowledge Protocols*. Indeed, some of the elements we find here, such as random [k] and the execution of blind signatures, and the last step of verification, $V = S/k$, performed by the receiver, utilize the logic that inspired zero-knowledge protocols.

Summary

In this chapter, we have analyzed hash functions, digital signatures, and blind signatures. After introducing hash functions, we started by describing the mathematical operations behind these one-way functions followed by an explanation of SHA-1. We then explained digital signatures with RSA and ElGamal with practical numerical examples and examined the possible vulnerabilities.

Finally, the blind signature protocol was introduced as a cryptographic instrument for implementing electronic voting and digital payment systems.

Therefore, you have now learned what a hash function is and how to implement it. You also know what digital signatures are, and in particular, you got familiar with the signature schemes in RSA and ELGamal. We also learned about the vulnerabilities that could lead to digital signatures being exposed, and how to repair them.

Finally, you have learned what blind signatures are useful for and their fields of application.

These topics are essential because we will use them abundantly in the following chapters of this book. They will be particularly useful in understanding the zero-knowledge protocols explained in *Chapter 5, Introduction to Zero-Knowledge Protocols*, and the other algorithms discussed in *Chapter 6, New Algorithms in Public/Private Key Cryptography*. Finally, *Chapter 6* will examine new methods of attack against digital signatures.

Now that you have learned the fundamentals of digital signatures, it is time to analyze zero-knowledge protocols in detail in the next chapter.

Section 3:
New Cryptography Algorithms and Protocols

In this section, we will describe the new protocols and algorithms that are emerging for data protection in the new environment of cybersecurity, blockchain, ICT, and quantum computing. Among these will be some patents and inventions of the author.

This section of the book comprises the following chapters:

5
Introduction to Zero-Knowledge Protocols

As we have already seen with the digital signature, the authentication problem is one of the most important, complicated, and intriguing challenges that cryptography is going to face in the near future. Imagine that you want to identify yourself to someone who doesn't know you online. First, you will be asked to provide your name, surname, and address; going deeper, you will be asked for your social security number and other sensitive data that identifies you. Of course, you know that exposing such data via the internet can be very dangerous because someone might steal your private information and use it for nefarious purposes.

Some time ago, I watched a video that impressed me. I have even decided to insert it into my presentations about privacy and security, and I presented it during an event related to Smart Cities in Silicon Valley where I had been invited to talk. The video starts with an alleged magician who invites people to enter a tent set up in the middle of the city. The magician reads, one by one, each person's past and something about their future. The unbelievable thing was that the magician (who had never met any of the people interviewed before) knew particulars about the astonished participants' lives only they were supposed to know.

How was it possible? Was it really magic or was it just a trick? At the end of the video, the trick was revealed. The magician knew everything about the participants' lives – finances, assets owned, and even credit card numbers – thanks to a staff of techno-hackers who sat behind a curtain, working hard to discover all the digital secrets they could about the participants. If a hacker knows your identity, they can easily find out about most of your digital life.

In another case, you might have read a news story where a gang of thieves planted a fake ATM in a commercial center. Each time a person inserted a card and entered their PIN, a computer recorded this information and the ATM refused the operation. Once the information had been collected from the cards and the PINs stolen, the hackers could then clone the cards, reproducing them along with their PINs, and subsequently be able to withdraw money at an actual ATM.

How is it possible to block this kind of scam? There are many situations in which sensitive information, such as passwords and other private information, is required. If a hacker obtains this information linked to a person or a machine, they can easily steal identities and wreak havoc for their victims.

One way to solve these kinds of problems is by not revealing any sensitive information, but this is not always possible. Another way is to avoid exposing private information by giving *proof of knowledge*. These cryptographic protocols are called **Zero-Knowledge Protocols** (**ZKPs**), which we are going to study in this chapter.

In this chapter, we are going to cover the following topics:

- Non-interactive and interactive ZKPs (the Schnorr protocol) with examples and possible attacks on them
- SNARK protocols and Zcash
- One-round ZKPs
- ZK13 and the zero-authentication protocol

Now that you have been introduced to the world of ZKPs and know what they are used for, it's time to go deeper to analyze the main scenarios and protocols used in cryptography.

The main scenario of a ZKP – the digital cave

Imagine this fantastic scenario: Peggy has to demonstrate to Victor that she is able to open a locked door in the middle of *Ali Baba's cave*, an annular cave with only one entrance/exit that can be reached from two directions, as you can see in the following figure. I suppose you have noticed that I have changed the names of the two actors, Peggy and Victor, from the usual Alice and Bob, just because here a verification is due, and the names **Peggy** and **Victor** match better with the first letters of **prover** (**P**) and **verifier** (**V**).

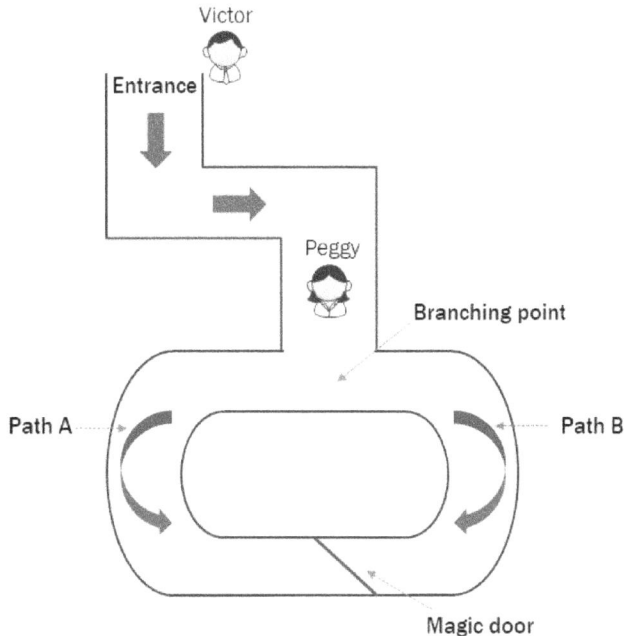

Figure 5.1 – Ali Baba's cave

This example involves Ali Baba's cave revisited in modern times, in which the door in the middle of the cave is locked through a secret electronic combination that's strong enough to prevent the entry of anyone who doesn't know the secret combination.

Now, suppose that Peggy must prove to Victor that she knows the combination to unlock the door without revealing the numbers to him.

So then, the challenge for Peggy is *not* to reveal the combination of the door to Victor, because Peggy is not sure whether Victor knows it and she doesn't want to give Victor any information about the combination. She just needs to demonstrate to Victor that she can exit from the opposite side of the cave.

Expressed differently, ZKP is a challenge where the answer is not revealing the exact information required, but simply proving to be able to solve the underlying problem. Indeed, the natural way to demonstrate knowing something is to reveal it, and the verification just comes naturally. However, by taking this approach of just demonstrating what you know so directly, you could reveal more information than required. With a ZKP, Peggy can avoid the issue of revealing the digital door's combination and, at the same time, Victor can be sure (even if he doesn't know the combination) that Peggy knows the combination if he sees Peggy coming out of the cave from the opposite side.

There are many ways to implement a ZKP because, as you can imagine, there are many scenarios in which such verification could be required.

For example, you can think of Peggy as a human and Victor as a machine (an ATM or server). Peggy has to identify herself to the machine, but she doesn't want to reveal any sensitive data, such as her name or surname. She just has to prove to the machine that she is really who she is supposed to be. The aim, in this case, is to avoid revealing Peggy's identity. ZKPs can be applied here. Another use case where ZKPs can be applied is the authentication of virtual machines in a computer network. We will cover this use case in *Chapter 9*, *Crypto Search Engine*, where we will use ZKPs to protect against man-in-the-middle attacks.

ZKPs can be applied to other kinds of challenges than authentication, such as the fields of nuclear disarmament and blockchain.

Now, we will delve deeper into the applications of ZKPs, starting with analyzing non-interactive protocols used to prove statements.

Non-interactive ZKPs

The protocol we are going to analyze in this section is a **non-interactive ZKP**. This means that the prover has to demonstrate the statement, assuming that the verifier does not know the solution (the content of the statement) and that the verification is made without any exchange of information from the verifier.

The scheme could be summed up as follows:

Prover (statement) —————-> [**Proof of knowledge**] —————> **Verifier** (verification)

Let's take this problem into consideration: Peggy states to know that a document [m] is encrypted with RSA, as follows:

```
m^e ≡ c (mod N)
```

Here, (N, c, e) are public parameters, and [m] is secret.

> **Important Note**
>
> Remember, I always denote the secret elements of the functions with the
> `[]` symbols.

In order to demonstrate the statement to Victor, the following protocol is executed:

1. Peggy chooses a random integer, `[r1]` (she keeps it secret), and calculates the inverse of `r1` (represented as `INV[r1]`) multiplied by `[m]` (modulo `N`):

   ```
   r2 ≡ [m] * r1^-1 (mod N)
   ```

2. Peggy calculates `(x1)` and `(x2)`, as follows:

   ```
   x1 ≡ r1^e (mod N)
   x2 ≡ r2^e (mod N)
   ```

   Peggy sends `x1` and `x2` to Victor.

3. Finally, Victor verifies the following:

   ```
   x1*x2 ≡ c (mod N)
   ```

It is supposed here that if Victor can verify *step 3*, `x1*x2 ≡ c (mod N)`, then Peggy really should know `[m]`.

As you can imagine, Peggy wants to demonstrate to Victor that she effectively knows the message `[m]` without revealing it. Remember that in this case, Peggy ignores whether Victor knows `[m]` or not.

So, the underlying challenge implicit to this statement is for Peggy to be able to solve the RSA problem without revealing `[m]`.

Indeed, it's supposed that if Peggy knows `[m]` (hidden in the cryptogram `(c)`), she can also calculate the function `x1*x2 ≡ c (mod N)`; otherwise, she will not be able to do that.

Another important consideration is the following: if `(c)` is a big number, it is supposed (I hope you will agree with me, but don't take this as gospel) that it should be hard to know `x1` and `x2` (the two factors of `(c)`) without knowing `[m]`. It could be considered the same degree of computational difficulty to factoring `(N)`.

As you can see in the preceding function, `(r2)` is calculated using `[m]` in the following equation:

```
r2 ≡ [m] * r1^-1 (mod N)
```

So, the final effect of the multiplication between $x1*x2$ (as you can see in the following demonstration) will be to eliminate $(r1)$ and leave m^e $(mod\ N)$, which is equal to (c).

Even if Victor doesn't know $[m]$, he can believe what Peggy states (to know $[m]$) because she has demonstrated she can *factorize* (c).

RSA is supported by the factorization problem (as we saw in *Chapter 3, Asymmetric Encryption*); here, the function is as follows:

```
x1*x2 ≡ c (mod N)
```

This states that it is computationally hard to find two numbers $(x1, x2)$ that factorize (c).

> **Important Note**
> I will prove that an attack on this protocol exists that avoids the factorization problem in order to trick Victor, which we will experiment with later in this chapter.

Numerical example:

Let's see with a numerical example how this protocol works before going deeper to analyze it:

```
m = 88
N = 2430101
e = 9007
m^e ≡ c (mod N)
88^9007 ≡ 160613 (mod 2430101)
```

Now, let's start the protocol of verification.

Peggy chooses a random number:

```
r1 = 67
```

- **Step 1**: Peggy calculates $r2$:

    ```
    r2 ≡ [m] * r1^-1 (mod N)
    ```

First, we calculate the `[Inv(r1)]` function (the inverse value of `r1` `(mod N)`), and then we multiplicate it for `[m]`:

```
67 * x ≡ 1 (mod 2430101)
x = 217621
[m]* x ≡ r2 (mod N)
88 * 217621 ≡ 2139941 (mod 2430101)
```

So, we have found `r2`:

```
r2 = 2139941
```

- **Step 2**: Peggy calculates `x1` and `x2`:

```
x1 ≡ r1^e (mod N)
x2 ≡ r2^e (mod N)
67^9007 = 1587671 (mod 2430101)
x1 = 1587671
2139941^9007 ≡ 374578 (mod 2430101)
x2 = 374578
```

Peggy sends `x1`, `x2` (`1587671`, `374578`) to Victor.

- **Step 3**: Finally, Victor can verify the following:

```
x1 * x2 ≡ c (mod N)
1587671 * 374578 ≡ 160613 (mod 2430101)
160613 = c (OK)
```

Let's see why this protocol is mathematically correct.

Demonstration of a non-interactive ZKP

We have to demonstrate the following:

```
x1 * x2 ≡ c (mod N)
```

Where `c ≡ [m]^e (mod N)`, we can substitute in the previous function (`x1 = r1^e`) and (`x2 = r2^e`) so that we get the following:

```
x1 * x2 ≡ r1^e * r2^e ≡ c (mod N)
```

Substituting r2 into the equation, we have the following:

```
x1 * x2 ≡ (r1)^e * (m * ((r1^-1)^e) ≡ (r1)^e * m^e * (Inv(r1)^e)
≡ c (mod N)
```

Going by the modular power's properties (collecting together the two factors I have highlighted), we have the following:

```
r1^e * (Inv(r1))^e ≡ 1 (mod N)
```

So, eliminating r1^e and (Inv(r1))^e from the final equation will leave only the m^e remaining in the second stage of the equation, and the result will be as follows:

```
x1 * x2 ≡ m^e ≡ c (mod N)
```

As you know, since the beginning of this demonstration, (m^e) is just the RSA encryption of the secret message [m], which is equal to the cryptogram (c); that is why x1*x2 = c. That's just what we wanted to demonstrate.

The next section will show how we can attack an RSA ZKP.

Demonstrating an attack on an RSA ZKP

If you have stayed with me until this point, I'm hoping you will follow me further on this journey, so I can give you a demonstration of using a protocol to trick the verifier.

Note that I created this attack at the end of 2018. This is one of the possible attacks on a ZKP that I have demonstrated.

The goal of this attack is to demonstrate that Eve (the attacker) can calculate two *fake* numbers (x1, x2) which prove to factorize (N) even if Eve effectively doesn't know [m].

Let's explore how this attack works and what effects are produced:

1. Eve (the attacker) picks up a random number, [r], and calculates the following:

    ```
    [r] * (v1) ≡ e (mod N)
    ```

    [r] and (e) are public, so that is known by everyone. By means of this function, Eve can extract (v1).

2. In parallel, Eve calculates the following:

    ```
    e * x ≡ c (mod N)
    ```

The parameters (e, c) are also known. This, just like in *step 1*, is an inverse multiplication (modulo N). The scope of this operation is to obtain [x]. Then, using [x], Eve multiplies [x] by [r], yielding (v2):

```
x * r ≡ v2 (mod N)
```

Eve sends (v1, v2) to Victor, who can verify the following:

```
v1 * v2 ≡ c (mod N)
```

Eve can impersonate Peggy, and she can claim to know [m] even if she doesn't know it!

Numerical example:

r = 39 is the secret number chosen by Eve.

(N, c, e) are the same public parameters of the previous example (N = 2430101, c = 160613, e = 9007).

- **Step 1**: Eve calculates v1:

```
e * r^-1 ≡ v1 (mod N)
9007 * 436172 = 1557988 (mod 2430101)
v1 = 1557988
```

- **Step 2**: Eve performs v2.

  The next operation is to obtain the inverse of (e) with respect to (c).

  e * x ≡ c (mod N), obtaining (x) in inverse modular multiplication:

```
9007 * x = 160613 (mod 2430101)
x = 2031892
```

  Then, using x, Eve obtains v2, performing the following operation:

  x * r ≡ v2 (mod N), obtaining v2:

```
v2 = 1480556
```

  After having gained v2, Eve sends (v1 = **1557988**; v2 = **1480556**) to Victor.

- **Step 3**: Verification stage.

  Finally, Victor can verify the following:

```
v1 * v2 = c
1557988 * 1480556 = 160613 (mod 2430101)
```

The attack was successful!

> **Important Note**
>
> Peggy herself could be the primary actor of this trick if she doesn't know [m],
> but she wishes to convince Victor of it.

This attack works because (c) contains [m], and I don't need to demonstrate showing
the value of [m]. This protocol isn't required to show [m] or its hash, [H(m)], because
Peggy doesn't want to reveal any information about [m] to Victor. Remember this is
not an authentication protocol, but it's a proof of statement (or knowledge) that Peggy
knows [m].

To use a hypothetical example, you can think of a scenario where there are two countries:
(A) has to demonstrate to (B) that it holds the formula for an atomic bomb. Using this
ZKP, (A) could claim to know [m] (the formula of the atomic bomb) without really
knowing it.

This attack could be avoided under one condition:

If Victor already knows [m], then he can require Peggy to send him a hash of the
message, H[m]. Victor can then verify whether (x1 and x2) are the correct values, and
he will accept or deny the verification based on the correspondence of the hash value
with [m].

In this case, the problem is that the aim of this protocol was not to prove something that
was already known but to prove something independently, regardless of whether or not it
was known.

This last point is very important because if Victor knows [m], then this protocol works; if
Victor doesn't know [m], this protocol fails.

To avoid this problem, we have to switch to an *interactive* protocol, as we will see in the
next section.

Schnorr's interactive ZKP

The protocol that we saw in the previous section is a non-interactive protocol, where
Peggy and Victor don't interact with each other but there is simply a *commitment* between
them. The commitment is that Peggy has to show that she knew the message [m] without
revealing anything about it. Thus, she tries to demonstrate to Victor that she can overcome
the RSA problem (or another hard mathematical problem) as proof of her honesty.
However, we have also seen that this protocol can be *bypassed* using a mathematical trick.

Let's see whether the following interactive ZKP is more robust and can prevent possibly
devastating attacks.

We always have Peggy and Victor as our main actors. So, let's assume the following:

- p is a big prime number.
- g is the generator of (Zp).
- B ≡ g^a (mod p) is the public parameter of Peggy.
- (p, g, B) are public parameters.
- [a] is the secret number object of the commitment.

Peggy claims that she knows [a]. In order to demonstrate the claim, Peggy and Victor apply the following protocol:

- **Step 1**: Peggy chooses a random integer, [k], where 1 ≤ k < p-1.

  She performs the following calculation:

  V ≡ g^**k** (mod p)

  Peggy sends (V) to Victor.
- **Step 2**: Victor chooses a random integer, (r), where 1 ≤ r < p-1.

  Victor sends (r) to Peggy.
- **Step 3**: Peggy calculates as follows:

  w ≡ (k - a*r) (mod p-1)

  Peggy sends (w) to Victor.
- **Step 4**: Finally, Victor verifies as follows:

  V ≡ g^w * B^r (mod p)

  If that is true, Victor should be convinced that Peggy knows [a].

Let's see why the protocol should work and the reason why the last function (V) states that Peggy really knows [a] (the commitment).

First of all, I will show why the protocol is mathematically true, and then I will give a numerical example of this protocol.

A demonstration of an interactive ZKP

Recall the following instructions:

```
V ≡ g^k  (mod p)
B ≡ g^a  (mod p)
w ≡ k - a*r  (mod p-1)
```

Now, we substitute all the past equations inside the last verification of *step 3*:

```
V ≡ g^w * B^r  (mod p)
V ≡ g^k  (mod p)
```

Substituting the functions in (V), the equation becomes the following:

```
V ≡ (g^ (k - a*r (mod p-1))) * ((g^a)^r)  (mod p)
```

For the properties of exponential factors, we have the following:

```
g^k ≡ g^ (k -[ar]) * g^[ar]  (mod p)
V ≡ g^k ≡ g ^(k -ar +ar)
```

Simplifying [-ar] with [+ar], we get back the following:

```
g^k ≡ g^k  (mod p)
```

That's what we wanted to demonstrate.

Let's do a numerical example to better visualize how this interactive ZKP works.

Numerical example:

```
p = 1987
a = 17
g = 3
```

(p = 1987 and g = 3) are public parameters.

[a] = 17 is the secret number that Peggy claims to know:

```
B ≡ g^a  (mod p)
```

(B) is the public key of Peggy, given by the following:

```
3^17 ≡ 1059  (mod 1987)
```

Peggy picks up a random number, `[k]` = 67, and she calculates (`V`):

```
V = 3^67 = 1753 (mod 1987)
```

Peggy sends (`V`) to Victor.

Victor picks up a random number (`r` = 37) and sends it to Peggy, who can calculate was following:

```
k - a * r ≡ w (mod p-1)
67 - 17 * 37 ≡ 1424 (mod 1987-1)
w = 1424
```

Peggy sends (w = 1424).

Finally, Victor now verifies whether (`V`) = 1753 corresponds to the following:

```
g^w * B^r ≡ V (mod p)
3^1424 * 1059^37 ≡ 1753 (mod 1984)
V = 1753
```

In fact, it does correspond.

Now, we analyze the reason why this protocol states that by knowing `[a]` automatically, Peggy can convince Victor. Let's use an example to better understand the problem.

This protocol can be used as an *authentication scheme* in which, for example, Victor is a bank that holds the public parameter (`B`) of Peggy (a client of the bank). The secret number `[a]` could be Peggy's secret code (PIN). In order to gain access to her online account, Peggy has to demonstrate that she knows `[a]`.

In another use case, we could have Victor as a central unit computational power (server) and Peggy as a user who wants to connect to the server using an insecure line, or again (as we will see later), Peggy could be another server, too.

The point of using a ZKP is to avoid Peggy revealing her sensitive data to the public. So, the underlying problem she has to demonstrate to Victor consists of solving a challenge in which she can demonstrate that she knows the discrete logarithm of (`B`).

As we have seen in *Chapter 3, Asymmetric Encryption*, knowing (`B`) and (`g`) is not enough to compute `[a]` in this function:

```
B ≡ g^[a] (mod p)
```

This is because we are operating in modular functions.

Of course, Peggy needs to know [a] if she wants to compute the verification function, (w):

```
w ≡ k - a*r (mod p-1)
```

There is no way to trick Victor, who moreover sent (r) to Peggy, which is used in the last verification function together with (v) and (B), along with (r), to be sure that Peggy cannot bluff:

```
V ≡ g^w * B^r (mod p)
```

So, I hope to have convinced you that there is no way for Peggy to trick Victor in this case.

Demonstrating an attack on an interactive ZKP

Now that we have seen that Peggy can't trick Victor, I have an attack against this protocol that I created in late 2018; let's see whether it works or not.

Here, the scope for an attacker (Eve) is to provide a final proof of verification without knowing the secret number, [a].

Step 1: Peggy chooses a random integer [k] where 1 ≤ k < p-1.

Then she calculates the following:

```
V ≡ g^[k] (mod p)
```

It's when Peggy sends (V) to Victor that Eve can try to shoot a man-in-the-middle attack.

Peggy sends (V) to Victor.

Step 2: Victor chooses a random integer, (r), where 1 ≤ r < p-1.

Eve injects V1 ≡ g^k1 (mod p), where [k1] is a number invented by Eve that substitutes [k].

Eve sends (V1) to Victor, substituting her value for Peggy's result. After receiving (r) from Victor, Eve calculates (v1) as follows:

```
V1 ≡ g^(v1) * B^r (mod p)
```

This is the path to the attack:

- If you can exclude (r) from the final verification function, (V1), then you have reached the goal.

- Essentially, the attacker should find a value for (v1), as follows:

```
v1 = [x] ---------> V1 = g^k1
```

Here are some notes:

- Remember that you don't have to implement (v1) in the same way the preceding function (v) did, but you are free to give (v1) any value.

- Remember that the earliest point of attack is substituting (V) with (V1), but that is not mandatory. In this case, the warning is that as you don't know (r) when you have delivered (V1) to Victor, this parameter can no longer be changed.

- Good luck! If you are able to find a way to trick the Schnorr interactive protocol, please let me know when you have arrived at a conclusion, and you will get a cryptographer researcher position.

The preceding analyzed interactive protocol suffers from another problem. Let's imagine that two people live in different time zones, such as Europe and Australia. If one is *ON*, the other one is probably *OFF* because they're sleeping. So, what happens if they have to wait for many hours to make or receive an economic transaction?

This protocol doesn't fit well with this kind of purpose, such as cryptocurrency transactions. Most cryptocurrency protocols use zero-knowledge algorithms to anonymize data inside their architecture structures. Now that we know how to implement such a protocol, we can explore zk-SNARKs.

An introduction to zk-SNARKs – spooky moon math

If you think ZKPs are pretty difficult to understand, that is because you haven't yet faced off with **zk-SNARKs** – a kind of ZKP, also known as **spooky moon math**. Here, the situation gets a little bit more complicated, but don't worry – it's not impossible. In the next section, you will see interesting new attack possibilities.

Non-interactive zero-knowledge proofs, also known as **zk-SNARKs** or **zk-STARKs**, are kinds of ZKPs that require no interaction between the prover and verifier, like the first protocol we saw in this chapter. In this section, we are going to focus on zk-SNARKs.

The name zk-SNARK stands for **Zero-Knowledge Succinct Non-Interactive Argument of Knowledge**. So, we are facing off with schemes that need only one interaction between the prover and the verifier.

Indeed, zk-SNARKs are very much appreciated for their ability to anonymize transactions and to identify users in cryptocurrency schemes, as we will see in this section.

The first cryptocurrency that adopted this new system to create *consensus* was **Zcash**.

The use of zk-SNARKs in a blockchain is important, as we will see later, for the use of smart contracts. As you may know, a smart contract is an escrow of cryptocurrency activated following the completion of an agreed execution.

Since smart contracts and blockchains are not a part of this book, I will show just a limited example of how zk-SNARKs work in a cryptocurrency environment, as it will be useful to understand.

For example, suppose Peggy makes a payment in Ethereum to execute a smart contract with Victor. In that case, both Peggy and Victor want to be sure that the execution of the smart contract (for Peggy) and the payment received (for Victor) are completed successfully. However, many details inherent to the smart contract will not be revealed. So, the role that zk-SNARKs play is fundamental to covering these secrets and executing smart contracts. In order to work, the protocol has to be fast, secure, and easy to implement.

As we have already seen, you will notice that this is just what the purpose of a ZKP is – to ease the navigation of an untrustworthy environment. Here, we are talking about blockchains and virtual payments, but essentially the process is similar.

So, in this environment, zk-SNARKs keep secrets by protecting the steps involved in a smart contract and, at the same time, proving that all these steps have been executed. This way, they protect the privacy of people and companies.

Remember that – not because you have to be super-skeptical, but because you should be realistic – this statement is true under determinate conditions, which I will try to explain as follows:

- The proof given by the prover holds the same computational degree of difficulty as the underlying algorithm chosen as a proof of knowledge.

- There is no mathematical way to trick the verifier with a shortcut or fake proof (such as substituting fake parameters into the $V1 \equiv g^\wedge(v1) * B^\wedge r \pmod{p}$ equation in order to avoid knowing [a]).

So, let's see how a zk-SNARK works.

Understanding how a zk-SNARK works

In this section, first of all, I will try to synthesize how zk-SNARKs generally work, and then we will return with a zk-SNARK protocol related to a proof of knowledge based on a discrete logarithm.

As we already have seen for the other ZKPs, a zk-SNARK is composed of three parts or items – (G), (P), and (V):

- G: This is a generator of keys, made by a private parameter (the statement or another random key) that generates public parameters given by private keys.
- P: This is a proof algorithm that states what the prover wants to demonstrate.
- V: This is a verification algorithm that returns a TRUE or FALSE Boolean variable from the verifier. I will demonstrate now that using ZKPs (and, in particular, zk-SNARK protocols) is *not* enough to keep [w] secret, but it is possible to arrive at proving the statement as TRUE if it is also FALSE.

Let's look at how a similar protocol example explained in the *Interactive ZKP* section (Schnorr) would work in a non-interactive way (zk-SNARK mode).

In this protocol, we have Anna as the prover and Carl as the verifier.

Here, Anna has to prove that [a] is known to her.

Anna calculates her private key, (y), given by the following:

```
y ≡ g^a (mod p)
```

(g) is a generator (as in D-H or other private-public algorithms we have already seen in this book).

Then, Anna picks up a random value, [v], inside p-1, which she keeps secret, and consequently, she can calculate the following:

```
t ≡ g^v (mod p)
```

Anna calculates (c) as a hash function of the three parameters, (g, y, t), and she can compute (r) as follows:

```
r ≡ v - c*a (mod p-1)
```

The verifier, Carl, can check the following:

```
t ≡ g^r * y^c (mod p)
```

Finally, if the verification validates the two terms of the function, then Carl accepts that the statement [a] proposed by Anna is TRUE.

Now that we have analyzed how this ZKP works in a zk-SNARK environment, let's see an attack on this protocol before we cover a numerical example.

Demonstrating an attack on a zk-SNARK protocol

This attack was performed by me in June 2019 and just goes to show that nothing is completely secure.

Let's say that Eve is an **Artificial Intelligence (AI)** server. We suppose that Eve (AI) intercepts the H(g, y, t) public hash function and performs a MiM attack.

While Peggy sends (V) to Victor, Eve substitutes (c) = H(g, y, t) with (c1) = H1(g, y, t1), remembering that (H) is the hash function and that (t1) is given by the following:

```
t1 ≡ g^v1 (mod p)
```

As you have probably noticed, substituting (v) with (v1) is the same trick that substituted (k) with (k1) in the previous attack.

Simply, Eve can put (r1) as follows:

```
r1 = v1
```

Now, Eve orders the AI (the intelligent third server connected to the internet) to send to Carl (r1, v1, c1), who can verify the following:

```
t1 ≡ g^r1 * y^c1 (mod p)
```

It's simple to demonstrate that t1 = g^v1 because of the following:

```
y^(p-1) ≡ 1 (mod p)
```

Finally, as we have assigned c1 = p-1 and r1 = v1, the final effect will be as follows:

```
t1 ≡ g^v1 ≡ g^v1 * 1 (mod p)
```

Numerical example:

```
p = 3571
g = 7
x = 23
```

Anna's public key is as follows:

```
y ≡ g^x (mod p)
7^ 23 = 907 (mod 3571)
y = 907
```

Now, I will show you how Anna can demonstrate to Carl to get the secret number, [x].

She chooses v = 67, as follows:

```
t ≡ g^v (mod p)
7^ 67 = 584 (mod 3571)
t = 584
```

Let's suppose that hash (g, y, t) as follows:

```
c = 37
```

She computes r as follows:

```
r ≡ v - c*x (mod p-1)
(67 - 37*23) ≡ 2786 (mod 3570-1)
r = 2786
```

Anna sends (r, t, c) = (2786, 584, 37) to Carl.

Carl can verify the following:

```
g^r * y^ c ≡ t (mod p-1)
7^2786 * 907^37 = 584 (mod 3571)
```

However, the AI intercepts the public parameters (y), (t), and (r). Eve leaves the (y) invariant, but she changes (t) with (t1) and (r) with (r1), performing a man-in-the-middle attack:

```
v1 = 57
```

Eve calculates the following:

```
t1 ≡ 7^57 (mod 3571)
t1 = 712
v1 = r1 = 57
c1 = p-1 = 3570
```

Eve sends `(r1, t1, c1) = (57, 712, 3570)` to Carl.

Carl verifies the following:

```
t1 ≡ g^r1 * y^c1 (mod p)
```

I highlighted the parameter that Eve substitutes, `(t1, r1, c1)`; she left the `(y, g, p)` invariant.

Substituting the new parameters into the equation of verification, we have the following:

```
7^57 * 907^3570 ≡ 712 (mod 3571)
```

Indeed, Carl is able to verify that `t1 = 712` corresponds with the parameters received from Eve.

Essentially, if Carl is not able to recognize that `r1 = v1` and/or he doesn't accept `c = p-1`, then the trick is done, and Eve can replace Anna.

So, what are the protections to adopt against this attack?

If this attack is implemented in a more sophisticated mode, it will probably be very difficult to avoid it.

Note that the parameter `(y)`, the public key of Anna that "envelopes" the private key `[a]` object of the statement, hasn't been modified during the attack.

Anyway, zk-SNARKs can be implemented using other methods and protocols to prove statements; we will see what these algorithms and protocols are in the next section. Blockchains and cryptocurrency are evolving quickly to find new methods to authenticate users anonymously. However, with this topic being relatively new, it is better to make the effort to find all the possible attacks and the repair methods for them.

Zk-SNARKs in Zcash cryptocurrency

In this section, we will analyze the zk-SNARK behind Zcash, a new cryptocurrency that aims to preserve the security and privacy of transactions, as discussed in a scientific paper released on November 12, 2020 (*Demystifying the Role of zk-SNARKs in Zcash*):

> *"The underlying principle of the Zcash algorithm is such that it delivers a full-fledged, ledger-based digital currency with strong privacy guarantees and the root of ensuring privacy lies fully on the construction of a proper zk-SNARK."*

A blog about Zcash stated the following regarding zero-knowledge proofs: *allow one party* (the prover) *to prove to another* (the verifier) *that a statement is true, without revealing any information beyond the validity of the statement itself.* In a version of zero-knowledge called "proof of knowledge," the prover can demonstrate a statement without revealing any sensitive information about the content.

But, as we have seen until now, these protocols are based on problems that are difficult to solve, such as factorization and discrete logarithms, and we have seen in the previous sections how some of them could be vulnerable to various kinds of attacks.

A reason why you should adopt a ZKP is to authenticate yourself without revealing too much information about your sensitive data or your server's identity.

Another reason for using a ZKP is to prove that you know *something* that you want to keep secret.

Here, we have two main different cases:

- In the first case, it is supposed that the verifier doesn't know the content of the statement.

- In the second case, it is supposed that the verifier already knows the content of the statement, and only has to verify its *correctness*.

All that becomes much more complicated if you decide to adopt a *non-interactive protocol* instead of an *interactive protocol* because, in the first case, the verifier does not need to give any further input. Indeed, as you can see, if a parameter is exchanged between the prover and the verifier, that could increase the security of the algorithm. On the other hand, we know that interactive protocols need many steps to reach the goal, hence they are not used as much as non-interactive protocols.

As we have seen in the previous sections, I have shown some attacks that are capable of deceiving the verifier, making them believe something that isn't true; moreover, if the verifier doesn't know the content of the statement, it is even *easier* to send fake mathematical parameters to trick them.

Now, the situation is changing fast in cryptography, for two reasons:

- The need for cryptography to address privacy and anonymize data for an increasing number of transactions over the internet and in e-commerce

- The rise in the use of digital payments and cryptocurrency

To anonymize exchanges of digital money and to ensure the privacy of users in cryptocurrency transactions, zk-SNARKs are adopted in several cases. The verifications used for the two main problems we have seen (RSA and the discrete logarithm) are almost obsolete. They are leaving room for new and more sophisticated, hard-to-solve problems; in Zcash, we see one such use case.

To get perfect zero-knowledge privacy, Zcash implemented a complex system of functions to determine the **validity of a transaction**.

In Zcash's consensus system, using zk-SNARKs, the objective is to return a response on the validity of a transaction without knowing anything about the content and terms of the transaction itself.

Zcash adopts a complex scheme to anonymize transactions and obtain a final answer as to whether transactions are valid. Here, I have tried to schematize the circuit:

Computation → Arithmetic circuit → R1CS → QAP → zk-SNARK

Computation	→	Arithmetic circuit	→	R1CS	→	QAP	→	zk-SNARK

Figure 5.2 – Circuit flow

Let's analyze what these functions represent one by one.

The first step is turning transaction validity into a mathematical function, which finally has to be gathered into a logical expression. This step is taken by creating an arithmetic circuit similar to a Boolean circuit (we saw this in *Chapter 2, Introduction to Symmetric Encryption*, and *Chapter 3, Asymmetric Encryption*) made by mathematical base operations (+, -, *, and /) and computed by Boolean operators (AND, OR, NOT, and XOR), so that the program converges into only one *gate*, which is the result of all the operations chained together, as we can see in the following example of computing the expression (a+b)*b*c:

Inputs   **Output**

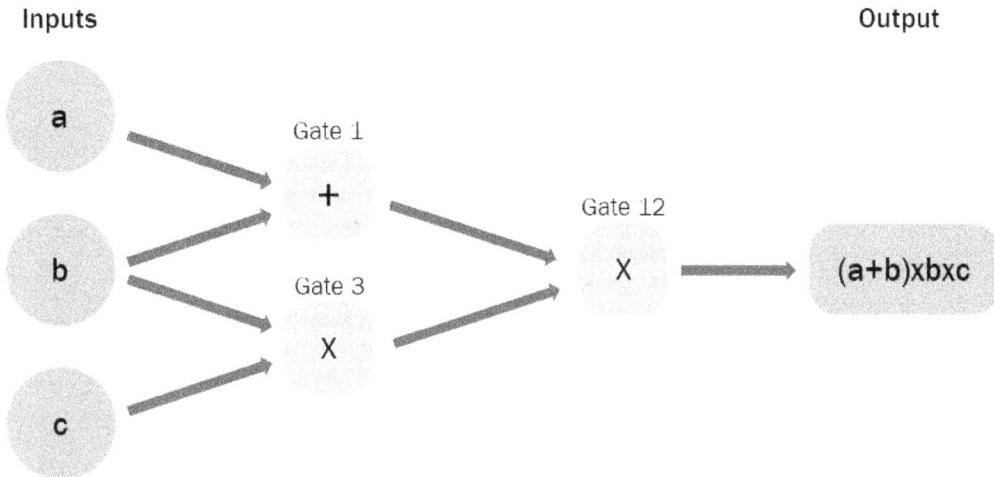

Figure 5.3 – An example of an arithmetic circuit. Inputs: a, b, and c. Output: (a+b)*b*c

As you can see, the **arithmetic** (or **algebraic**) circuit converges in only a single gate all the operations performed on the inputs: a, b, and c. Looking at this circuit, from left to right, we have the single terms (**a, b**, and **c**); first, (a) is added to (b), then (b) is multiplied by (c), and finally, the result is multiplied by the result of the previous sum. All that is mathematically expressed in only one final gate. We can represent layers and layers of these operations, reducing them all to one arithmetic circuit.

The second step is a **Rank 1 Constraint System (R1CS)** representation.

In R1CS, we have a group of three vectors (**A**, **B**, and **C**), as you can see in the following figure. The solution to satisfy the system is a new vector (S) given by an operation of a (.) dot product between the vectors, of which the final result has to be zero. So, R1CS has this scheme of operations and must satisfy the following equation with a result of zero:

```
(S • A) * (S • B) - (S • C) = 0
```

For example, this is a satisfied R1CS system:

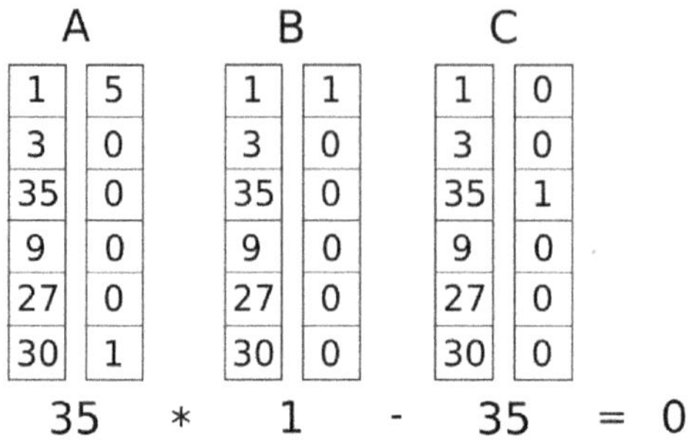

Figure 5.4 – An example of a satisfied R1CS system

As you can see in *Figure 5.4*, the value of the vector [S] is [1, 3, 35, 9, 27, 30], which ensures a satisfied R1CS system.

Indeed, if you look at column [A], the result of the operations (.) at the end of the column is as follows:

```
[1.5 + 3.0 + 35.0 + 9.0 + 27.0 + 30.1] = [5 + 0 + 0 + 0 + 0 +
30] = [35]
```

The result of the operations in column [B] is as follows:

```
[1.1 + 3.0 + 35.0 + 9.0 + 27.0 + 30.0] = [1 + 0 + 0 + 0 + 0 +
0] = [1]
```

Finally, the result of the operations in column [C] is as follows:

```
[1.0 + 3.0 + 35.1 + 9.0 + 27.0 + 30.0] = [0 + 0 + 35 + 0 + 0 +
0] = [35]
```

Therefore, R1CS checks whether the values are *traveling correctly*. It's a verification of the values. In our example in *Figure 5.4* for instance, R1CS will confirm that the value coming out of the multiplication gate where (b) and (c) went in is (b*c).

In the third step, Zcash converts R1CS *flat code* to a **Quadratic Arithmetic Program** (**QAP**), which operates on polynomials in (mod x).

So, the next step is taking R1CS and converting it into QAP form, which implements the same logic as before, except using polynomials instead of dot (.) products between vectors.

As I told you before, I will limit myself to explaining the Zcash process at a high level, so I will not go any deeper in analyzing the third step of the QAP.

At this point, can you guess why the inventors of Zcash put so much effort into this system?

It is probably because the inventors aspired to create the *perfect* ZKP. Indeed, in the paper entitled *Aurora: Transparent Succinct Arguments for R1CS*, the authors stated that their goal is to obtain transparent zk-SNARKs that satisfy the following conditions:

- **Post-quantum security**: *This is motivated by the desire to ensure the long-term security of deployed systems and protocols.*

- **Concrete efficiency**: *We seek argument systems that not only exhibit good asymptotics (in argument size and prover/verifier time), but also demonstrably offer good efficiency via a prototype.*

Given the high expectations predicted by the authors for this protocol, let's go on to explore the final steps of the sequence.

This protocol offers a probabilistic solution by performing a multiplication of polynomials. If the two polynomials match at a random point, we can be confident that the chosen point verifies the proof correctly. The reason for this transformation is that instead of checking the constraints individually, as in R1CS, we can now check all the constraints at the same time. Here, you can see an example of how the vector verification looks in a QAP:

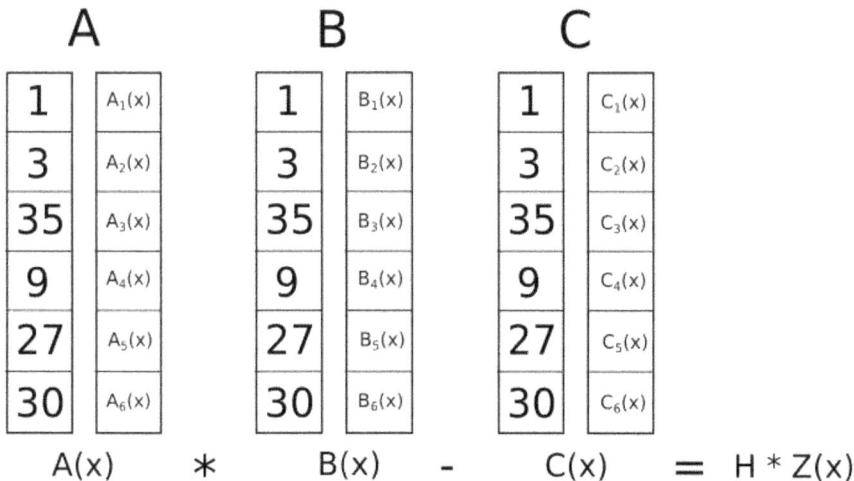

Figure 5.5 – An example of the vector verification in a QAP

Can you recognize the differences between this representation and the *checksum* in R1CS?

In both cases, if the logic gate is equal to zero, the result given by the dot (.) product of the checks passes; if at least one of the (x) coordinates gives a non-zero result, this means that the values going into and out of that logic gate are inconsistent.

But at this point, there could be a problem: if someone knows in advance which point the verifier chooses to check the validity, they can generate an invalid polynomial, but it could still satisfy that point.

Essentially, this is a dangerous step and could be vulnerable to attacks. To overcome this problem, Zcash applied sophisticated techniques to zk-SNARKs in order to evaluate the polynomials *blindly*. These mathematical techniques used in Zcash, such as homomorphic encryption and pairings of elliptic curves, help to blind the operations but increase complexity. We will look at these issues in the next section, but now we will finish discussing the entire process of zk-SNARKs in Zcash.

As I said at the beginning of the Zcash explanation, the goal of the protocol is to determine whether a statement (in this case, a transaction) is true or false, thereby preventing the *double-spending* problem.

Conclusions and the weaknesses of zk-SNARKs in Zcash

As we saw in the previous section, one of the weak points of this protocol can be found in QAP. As I have explained, Zcash has tried to overcome this problem using homomorphic evaluation, in other words, keeping the polynomials *in blind*. The issue is that homomorphic encryption usually causes bit-overflow; moreover, the protocols and schemes required to achieve fully homomorphic encryption are very complex. As you already know, my theory is that *in cryptography, complexity is the enemy of security*. I won't enter this debate now because it's not within the scope of the book to analyze the entire protocol of Zcash.

Imagine the scenario discussed in the *Non-interactive ZKPs* section based on the RSA problem. If I have to demonstrate to an expert that I hold the formula for an atomic bomb, then the experts will probably ask me to show something more than a hash function of the document, [m], that states the proof. The verifier will be convinced only when they get substantial proof. In other words, ZKPs are limited in the amount of evidence of knowledge they are able to provide.

One-round ZKP

In this section, we'll explore a little-known ZKP composed of only *one round* of encryption that was presented by two researchers, Sultan Almuhammadi and Clifford Neuman, of the University of Southern California, which purports to give proof of knowledge for a challenge in just one round. The paper states the following: *"The proposed approach creates new protocols that allow the prover to prove knowledge of a secret without revealing it."*

The researchers also proved that a non-interactive ZKP is more efficient in terms of computational and communications costs because it saves execution time and reduces latency in communication.

ZKPs are used in many fields of information technology, such as e-commerce applications, smart cards, digital cash, anonymous communication, and electronic voting. Almuhammadi and Neuman sought to satisfy the requirements of ZKPs but in just one round, eliminating any iterative mathematical scheme that would entail high computation and communication costs.

So, let's dive deep to analyze this one-round ZKP and see how it works.

Let's say that Peggy wants to demonstrate to Victor that she knows a discrete logarithm (we'll be focusing on a discrete logarithm, but the protocol can work for other problems); in order to do this, Peggy has to demonstrate that she knows `[x]`, as follows:

```
g^[x] ≡ b (mod p)
```

Victor launches a challenge (c) to verify whether Peggy really knows `[x]`. He picks up a random `[y]` and calculates the preceding function:

```
c ≡ g^[y] (mod p)
```

Victor sends (c) to Peggy. She inserts the parameter `[x]` on (c), computing (r) as follows:

```
c^[x] ≡ r (mod p)
```

Peggy sends (r) to Victor, who can verify the following:

```
r ≡ b^[y] (mod p)
```

Finally, Victor accepts the verification if (r) corresponds to `V = b^[y] (mod p)`.

This protocol looks very simple and straightforward. It is based on the computational difficulty to calculate the discrete logarithm, as you have seen in previous cases. But to help you better understand the operations, I will show how it works mathematically and demonstrate a numerical example in the next section.

How it works mathematically

The first question is: why are the parameters (r and V) mathematically identical?

Here, you can find the answer:

```
r ≡ c^[x] ≡ g^[y]^[x] ≡ g^[y*x]  (mod p)
V ≡ b^[y] ≡ g^[x]^[y] ≡ g^[x*y]  (mod p)
```

As you can see, $r \equiv V \equiv b^{\wedge}y \pmod{p}$.

Numerical example

Let's look at a numerical example:

```
p = 2741
g = 7
```

$x = 88$ is the secret number that Peggy has to demonstrate that she knows.

The statement is as follows:

```
g^x ≡ b  (mod p)
7^88 ≡ 1095  (mod 2741)
b = 1095
```

Victor chooses a random number:

```
y = 67
```

Victor calculates the following:

```
g^y ≡ c  (mod p)]
7^67 ≡ 1298  (mod 2741)
c = 1298
```

Peggy, after having received (c), calculates the following:

```
c^x ≡ r  (mod p)
 1298^88 ≡ 361  (mod 2741)
 r = 361
```

Peggy sends (r) to Victor, who can verify the following:

```
b^y ≡ V  (mod p)
 1095^67 ≡ 361  (mod 2741)
 V = 361 = r
```

Finally, it is verified!

As you can see, we have proved the one-round ZKP with the help of a numerical example. In the next section, I will demonstrate the strong similarity of this protocol with a protocol we have analyzed before in this book: **Diffie-Hellmann (D-H)**.

Notes on the one-round protocol

Having analyzed this protocol, you may have noticed that it is similar to the D-H exchange. Undoubtedly, the authors of the one-round ZKP were well aware of that. Still, even though the aim of the one-round ZKP is different from that of D-H, I will compare the two algorithms so that you can see what similarities there are.

In the second part of the analysis, we will see how efficient this protocol is. Indeed, with only two steps, Peggy can demonstrate to Victor that the statement [x] is valid.

Now, we can reassemble the one-round protocol using the following method:

Step 1: Peggy

```
g^x ≡ b  (mod p)
```

That is the same in D-H as the following:

```
g^a ≡ A  (mod p)
```

Step 2: Victor

```
g^y ≡ c  (mod p)
```

That is the same in D-H as the following:

```
g^b ≡ B  (mod p)
```

Step 3: Peggy

```
c^x ≡ r (mod p)
```

In D-H, this becomes the shared key H:

```
B^a ≡ H (mod p)
```

Step 4: Victor

```
b^y ≡ r (mod p)
```

In D-H, this again becomes the shared key H:

```
A^b ≡ H (mod p)
```

So, [H] is the shared private key that "Alice and Bob" (here, Peggy and Victor) use to compute in D-H. Here, it is just [r = H] that gives the proof to Victor.

So, we can certainly say that Sultan and Clifford's protocol is identical to D-H, as discussed in *Chapter 3, Asymmetric Encryption*.

This protocol undoubtedly verifies that Peggy knows [x]. She can demonstrate it to Victor even if Victor doesn't know [x]. That is the exciting point, and the innovation of this protocol: even if Victor doesn't know [x], by using this protocol, he can be confident that Peggy knows it. In other words, what the authors of this protocol did was apply the D-H protocol to the ZKP use case.

If you look at the simplified version of the protocol shown below, you will get an even better understanding of the steps required. There are only two, essentially:

- Initialization of the parameters for Peggy is g, b, p, and x.
- Victor generates a random y.

Step 1: Victor sends the following to Peggy:

```
c ≡ g^y (mod p)
```

Step 2: Peggy sends the following to Victor:

```
r ≡ c^x (mod p)
```

Instantly, Victor can verify the following:

```
r ≡ b^y (mod p)
```

As you can see, there are only two steps required to perform this protocol and verify the statement [x] through Victor's validation of the parameter (r).

This protocol gave me the inspiration to build a new protocol, which we will explore in the next section. My research has allowed me to reduce the number of steps to one.

ZK13 – a ZKP for authentication and key exchange

The **ZK13** protocol was invented and patented by me in 2013. It's a non-interactive protocol that solves an issue that was very important to my **Crypto Search Engine** (CSE) project: authentication without a public key.

In this section, we will analyze this ZKP that's used for authentication. It doesn't matter whether it's humans or computers; we could call this protocol a **ZK-proof of authentication**. To better understand the problem, imagine Alice and Bob want to share a common secret, something that only they know. Let's say that the secret is the answer to the following question: how many birds were counted at the lake shoreline today? The answer is known only to Alice and Bob, unless they have revealed it to someone, but this is a problem we will take into consideration later. For now, nobody else can know the answer except Alice and Bob. We can consider the number of birds counted as a shared secret, a key that doesn't need to be exchanged. It is shared by Alice and Bob, a key that is implicitly formed by a common experience. So, besides the authentication problem, there is also the problem of verifying a private **Pre-Shared Key** (PSK). Indeed, under ZK13, Alice tells Bob to use the secret shared key (the number of birds counted) as a secret password, [private key]. What's even more interesting here (and this is where it really differs from the D-H key exchange algorithm we saw before) is that the secret key is not *really* exchanged at all, but instead is simply something that is known to both parties and is only verified.

So, the problems that this ZKP can solve are several. Here, we will just consider the authentication problem. Later in the book, we will analyze how to use a ZKP to exchange a shared private key.

In 2013, I was drawing up the architecture of the CSE, a project that has absorbed a large part of my professional life. We will talk in detail about the CSE in *Chapter 9, Crypto Search Engine*. At the time, one of the toughest problems to solve with the CSE's architecture was finding a cryptographic method to identify the **Virtual Machines** (**VMs**) network. Since the algorithm chosen was symmetric, the problem was to find a method of authentication that would work with the symmetric algorithm. As you already know from *Chapter 2, Introduction to Symmetric Encryption*, it's common to think that symmetric algorithms don't have a digital signature method of authentication because they do not have public keys. At first glance, it doesn't seem easy to find such a method of authentication, but it can be possible if the process starts with a shared secret. The goal was to prevent a man-in-the-middle attack by an external hostile VM against a network.

To overcome all these issues, I considered implementing a new ZKP. Taking a look at the most popular ZKPs, I did consider Schnorr (presented earlier in this chapter) as a candidate. But an *interactive protocol* didn't fit well. This scheme needs more steps between the prover and verifier, generating latency in the communication. So, I decided to implement a new *personal* zero-knowledge non-interactive protocol.

After many studies and a pinch of inventiveness, I designed ZK13. Before analyzing it, I will explain what constraints I worked under:

- The secret shared key (the challenge) has to be embedded inside the VM database. Therefore, engineers could *inject* the secret parameter [s] into both of the VMs without exchanging any keys through an asymmetric algorithm.

- The goal of ZK13 is to enable parties to identify each other by sharing a small amount of sensitive information. This means exchanging only the minimum amount of sensitive information (that is, the hash of [m]: H(m) instead of [m] itself) that needs to be shared. Indeed, the greater the amount of information exchanged, the greater the vulnerability to attack becomes.

- ZK13 had to be a simple and, as I have already said, *non-interactive* protocol. Therefore, only one piece of information should be required by the prover. The reasons for this are twofold: first, to avoid an excess of information being exchanged (see the previous point) because that could compromise security. The second reason is related to the goal of the application: the CSE is a platform on which encrypted data is searched and retrieved using the cloud or external servers. Because a search engine has to be fast, queries should be fielded and answers given in the least amount of time possible. So, it is crucial to avoid latency during the authentication phase.

- Another constraint of ZK13 was for it to use the best and most secure authentication methods. At the time that it was conceived (2011–2013), the quantum computing era was not yet seen as dangerous for cryptography. So, the underlying problem on which the system relied was the discrete logarithm, which is still considered a hard problem.

ZK13 explained

The ZK13 protocol, with only one transmission and a shared secret, is presented as follows:

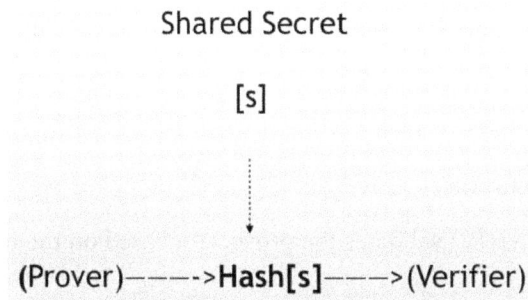

Shared Secret

[s]

(Prover)––––->Hash[s]––––>(Verifier)

Figure 5.6 – The scheme of the shared hash[s] secret

Let's dive deeper into ZK13 and see how it works.

Bob (VM-1) has to prove that he knows the secret, [s], to Alice (VM-2) in order to send Alice a set of encrypted files using the CSE system. Remember that [s] is stored inside the *brains* of both Alice and Bob, the two VMs, as an innate native injected secret.

Bob picks a random number [k] (the (G) element of a zk-SNARK or the random key generator). This random number, [k], is generated and destroyed in each session:

Public parameters:

p: This is a prime number.

g: This is the generator.

Key initialization:

[k]: This is Bob's random key.

Secret parameters:

[s]: This is the common shared secret.

H[s]: This is the hash of the secret, [s].

Step 1a: Bob calculates (r) as follows:

```
r ≡ gᵏ (mod p)
```

Let's say that the secret shared is `[s]`, but effectively, the VM operates with `H[s]`, the hash functions of `[s]`.

Step 1b: Bob calculates `[F]`, a secret parameter, which is changed in each session (just because `[k]` changes):

```
H[s]*k ≡ F (mod p-1)
```

Now, with (g) raised to `[F]`, Bob proves (P), which is the second element of the zk-SNARK:

```
gᶠ ≡ P (mod p)
```

Bob sends the pair (P, r) to Alice.

Step 2: The verification step (V) validates the prove, (P), based on the function:

```
[s] ──> H[s]
r[Hs] ≡ gᶠ (mod p)
```

If:

```
V ≡ r[Hs] = P (mod p)
```

Alice proceeds to make a hash of `[s]` : `H[s]`, and then she accepts the authentication if `V = P`; if ((V) is not equal to (P)), she doesn't accept the validation.

In this case, as we have supposed in the initial conditions, `[s]` is supposed to be known by Alice.

As you can see, ZK13 works in only two steps, but the verifier (in this case, Alice) must know the secret, `[s]`; otherwise, it is impossible to verify the proof.

Numerical example:

Now, let's see a numerical example of the ZK13 protocol:

Public parameters:

```
p = 2741
g = 7
```

Secret parameters:

```
H[s] = 88
k = 23
g^k ≡ r (mod p)
7^23 ≡ 2379 (mod 2741)
r = 2379
```

Now, Bob calculates [F] and then (P):

```
[Hs] * k ≡ F (mod p-1)
88 * 23 ≡ 2024 (mod 2741-1)
F = 2024
g^F ≡ P (mod 2741)
7^2024 ≡ 132 (mod 2741)
P = 132
```

Alice verifies the following:

```
r^[Hs] ≡ P (mod p)
2379^88 ≡ 132 (mod 2741)
```

Alice double-checks whether [Hs] = [s]; if it's TRUE, then it means that Bob does know the secret, [s]. Now that we have proven that ZK13 works with a numerical example, I want to demonstrate how it works mathematically.

Demonstrating the ZK13 protocol

Since P ≡ g^F (mod p), what we want to demonstrate is the following:

```
P ≡ g^F ≡ r^s (mod p)
```

(Here, I use [s] for the demonstration instead of H[s].)

As r ≡ g^k (mod p), substituting (r) in the preceding equation, we have the following:

```
P ≡ g^F ≡ (g^k)^s (mod p)
```

We also know that F is the following:

```
F ≡ s*k (mod p-1)
```

Finally, for the properties of the modular powers substituting both `[F]` and `(r)`, we get the following:

```
P ≡ g^s*k ≡ (g^^k)^^s ≡ g^^k*s (mod p)
```

Basically, I have substituted the parameter `(P)`, the proof created by Bob, with the elements of the parameter itself, demonstrating that the secret, `[s]`, is contained inside `(P)`. So, `(P)` has to match with the *ephemeral* parameter `(r) ^ [s]` generated by Bob and sent to Alice together with the proof, `(P)`. If Alice knows `[s]`, then she can be sure that Bob also knows `[s]` because `(P)` contains `[s]`. That's what we wanted to demonstrate.

Notes and possible attacks on the ZK13 protocol

You will agree with me that using this protocol, it is possible to determine proof of knowledge of the secret, `[s]`, in only one transmission.

During the explanation of the algorithm, I have divided it into three steps, but actually, there are only two steps (with only one transmission), because the operations of `(G)` key generation are offline. So, steps 1a and 1b can be combined into effectively only one step.

Possible attacks on ZK13

Let's say Eve (an attacker) wants to substitute herself for Alice or Bob, creating a man-in-the-middle attack.

This could be done as follows.

Eve replaces `(r)` with `(r1)`, generating a fake `(k1)`, by calculating the following:

```
r1 ≡ g^k1 (mod p)
```

But when Eve computes `[F]`, she doesn't know `H[s]` (because it's assumed that `[s]` will remain secret). So, this attack fails.

Instead, she can collect `(r, P)` and replay these parameters in the next session, activating a so-called **replay attack**.

This attack could be avoided here, because `(r)` is generated by a random `[k]`, so it is possible to avoid accepting an `(r)` already presented in a previous selection.

So, that was one attack that could be faced, and we saw how to avoid it.

Summary

Now you have a clear understanding of what ZKPs are and what they are used for.

In this chapter, we have analyzed in detail the different kinds of ZKPs, both interactive and non-interactive. Among these protocols, we saw a ZKP that used RSA as an underlying problem, and I proposed an original way to trick it.

Then, we saw the Schnorr protocol implemented in an interactive way for authentication, on which I have proposed an attempt to attack.

Moving on, we explored the zk-SNARKs protocols and *spooky moon math*, just to look at the complexity of some other problems. Among them, we saw an interesting way to attack a discrete logarithm-based zk-SNARK. We dived deep into Zcash and its protocols to see how to anonymize the transactions of this cryptocurrency.

Later in the chapter, we encountered and analyzed a non-interactive protocol based on the D-H algorithm. Finally, we explored ZK13, a non-interactive protocol, and its use of shared secrets to enable the authentication of VMs.

You became familiar with some schemes of attacks, such as man-in-the-middle, and used some mathematical tricks to experiment with ZKPs.

The topics covered in this chapter should have helped you understand ZKPs in greater depth, and you should now be more familiar with their functions. We will see in later chapters many links back to what we explored here. Now that you have learned the fundamentals of ZKPs, in the next chapter, we will analyze some private/public key algorithms that I have invented.

6
New Algorithms in Public/Private Key Cryptography

In 2001, I met a person who played an important role in my professional life; his pseudonym is **Terenzio**.

It was a sunny summer day; Terenzio was going back and forth running toward a tube slide and then into a swimming pool. I was reading a journal in the shade when I heard a particular voice that repeated some strange mathematical formulas. Time after time, Terenzio passed near me after jumping into the swimming pool. Terenzio is an autistic guy, and he suffers from dyslexia, which affects his speech. However, mother nature gave him a big gift; he can literally see big numbers in his mind, and in particular, prime numbers. We became friends, and I investigated his formulas. I have to be grateful to Terenzio even though I am writing this book on cryptography. The inventions I have created during my 20-year career are dedicated to, and in part are the merit of, Terenzio.

In *Chapter 5*, *Introduction to Zero-Knowledge Protocols*, I explained my non-interactive protocol that was patented in 2013: ZK13. In this chapter, I will explain two algorithms that were patented and published between 2008 and 2012. The two algorithms we will talk about are called MB09 and MBXI – these are my initials combined with the year of their invention. Additionally, we will look at a third protocol, MBXX, patented in 2020, which is a union between MB09 and MBXI from a new perspective that is related to blockchain and digital currency. Finally, I will introduce the concept of one of my new patents: **brain password**. This allows a computer to identify a human via brain waves. The system is still in process, with a demo set up to show that it works. Through minimal hardware (similar to headphones), brain password is able to detect brain waves, using them as biometric authentication to unlock a device or any kind of application. However, the objective of this chapter are the algorithms of cryptography invented in my career, so let's focus on them.

We will begin with an introduction to the genesis of the algorithms. Following this, we will examine, in detail, the scheme of the algorithms, their strengths, and their weak points.

We will start with MB09, which is a public/private key algorithm concept used principally for digital money transmission. Moreover, we will see the digital signature schemes of MBXI and a comparison with **Diffie-Hellmann (D-H)** and RSA.

Finally, we will discuss one of the last cryptographic protocols, MBXX, which could be considered valid for the blockchain consensus.

In this chapter, we will cover the following topics:

- The genesis of the MB09 algorithm
- The scheme and explanation of MB09
- A detailed explanation of MBXI
- Unconventional attacks on RSA
- Digital signatures in MBXI
- MBXX – the evolution of MB09 and MBXI in the light of the blockchain revolution and the consensus problem

Several of these algorithms are related to the *digital payment* environment, so we will dedicate a part of this chapter to discussing this environment, touching on blockchain and cryptocurrency.

Now, let's look at the genesis of the first algorithm, MB09, and its related applications.

The genesis of the MB09 algorithm

When I started project MB09 and, consequently, MBXI, at the time, I didn't know which applications they would be suitable for. In fact, I didn't know whether they would have any application in cryptography or cybersecurity.

Between 2007 and 2008, I decided to start applying for my first patent, the system that later I called MB09. At the time, I was studying some problems related to digital payments, and the goal was to implement an algorithm that could fit this kind of environment. When I finally completed the application for MB09, it was June 12, 2009. It is likely that at that time, only a few people knew about a new wave in *crypto-finance* guided by a group of cyber-punks and crypto-anarchists. This new wave was rising as one of the most innovative technologies invented in the 21st century.

With their headquarters in an artistic, funky building inside a pseudo-hotel located at 20 Mission Street in the heart of San Francisco, the newly born group set up a *hacker house* to meet and discuss the new technofinance. When I visited this place some years ago, the group had already left it, but the stories of epic brainstorming and meetups created a sense of mystery around this phenomenon.

Figure 6.1 – The hacker house at 20 Mission Street, San Francisco – headquarters of Bitcoin (photograph by the author)

Among this group, there was probably a legendary figure called Satoshi Nakamoto, the mysterious author of a published white paper, titled *A Peer-to-Peer Electronic Cash System*, on digital payments. It was November 2008. That was also the beginning of a new era, which was destined to forever change the digital payment system.

In the abstract of his white paper, Satoshi stated the following:

> *"We propose a solution to the double-spending problem using a peer-to-peer*
> *network. The network timestamps transactions by hashing them into an*
> *ongoing chain of hash-based proof-of-work, forming a record that cannot*
> *be changed without redoing the proof-of-work. The longest chain not only*
> *serves as proof of the sequence of events witnessed, but proof that it came*
> *from the largest pool of CPU power."*

The paper never mentioned the name *blockchain*, but it is evident that it refers to it when it says *the longest chain*.

It was the prelude of the age of cryptocurrency, which culminated in the rise and common usage of one of the most extraordinary inventions in virtual money: Bitcoin. Nowadays, who hasn't heard about Bitcoin? Perhaps we know a little about Zcash (we discussed this in *Chapter 5, Introduction to Zero-Knowledge Protocols*) and all the myriads of virtual coins released in the period of **Initial Coin Offering** (**ICO**). It was a correlated phenomenon linked to Bitcoin, something similar to an **Initial Public Offering** (**IPO**) but made up of cryptocurrency virtual tokens, starting from 2016.

This chapter will focus on the algorithms (that is, those related to secure and anonymous payments) that were invented in the period between 2008 and 2012. It's essential to understand the context in which the idea of a new, free, and anarchic way of conceiving money has arisen. Additionally, it's important to understand the technology behind it: the blockchain. It was probably the fulcrum and the core of digital payment systems. Now, it's rising as one of the emergent technologies that enables many applications in FinTech, artificial intelligence, healthcare, and other sectors.

However, we have to go back a few years in this story. As you will probably remember, I told you about David Chaum (*Chapter 4, Introducing Hash Functions and Digital Signatures*) and his DigiCash, which was crafted in the 1990s. Blind signatures anonymized this form of digital payment. It was undoubtedly the first attempt to create digital money, but it wasn't enough to fight against the finance colossus that wanted to maintain the authority of financial power. In 2008, and precisely on the Wednesday after September 15 of that year, when financial colossus Lehman Brothers collapsed, many other banks were also involved in bankruptcy and bailouts.

An extraordinary financial crisis called the **Subprime Mortgage Crisis** laid the foundations for this new wave of cryptocurrency. The collapse of the financial system and people's deep uncertainty pushed Satoshi Nakamoto to re-think the way to produce, transmit, and spend money. Indeed, in his already mentioned white paper, the three pillars predicted by Satoshi Nakamoto to build the perfect currency are *no government*, *no banks*, and *no trusted third parties*. On January 3, 2009, when Satoshi crafted the first *genesis block* of Bitcoin, he sculpted in the *blockchain* a note next to the numbers generated by the first hash of the chain. This note read *Chancellor on brink of second bailout for banks* and remarked on his hatred of banks and the traditional financial system.

In the following screenshot, you can see the genesis block of Bitcoin as it appeared:

Figure 6.2 – The genesis block by Satoshi Nakamoto

Analyzing the blockchain and the cryptography behind it is beyond the scope of this book. However, what was difficult for Satoshi Nakamoto was finding a way to perform the validation process for the transmission of digital currency, which is called the **consensus problem**. I will return to this concept of *consensus* later in this chapter to demonstrate that it is possible to avoid it, under certain conditions, using only the power of cryptography algorithms.

Besides all of these events (in fact, I was not yet conscious at that time of this revolution), two other motivations pushed me to project algorithms in the field of public/private key encryption. The first was related to another payment method, called **M-Pesa**, which I came across during my research (2007–2008). The second was related to my big wish to try to fight against the consolidated standard algorithms.

Let's look at the first motivation. I suppose that not many of you know what M-Pesa is and how it works. It is a straightforward payment method, so I will not spend more than a few words explaining its mechanism. The interesting question is why do two-thirds of African people (more than 1.2 billion) use M-Pesa to send money via cellphones and receive the majority of the $500 billion from emigrants in other countries?

The answer is equally simple: low transaction fees (compared to MoneyGram and Western Union's higher fees) and the simple method of using the application on a phone. It's well known that in Africa there are more cellphones than cars, houses, or bank accounts. Africa will be the next colossal market for finance; distances are impossible to cover with bank branches or ATMs to withdraw money. So, the most widely used method of payment and transmission of funds is using a phone. This phenomenon attracted so much of my attention that I spent months investigating the systems of payment adapted for cellphones.

At the time, Vodafone (the big telecommunication company) owned 40% of Safaricom (Kenya's biggest telecom company). In 2007, Safaricom began a pilot program that allowed users to send money via cellphone. Vodafone had rolled out the product in Tanzania, South Africa, Mozambique, Egypt, Fiji, India, and Romania, beginning with Kenya itself.

In May 2011, after having implemented my first crypto-digital payment system, showing that it worked, I went to sign a contract with a big telecommunication company involved in digital payments (I have mentioned one of these previously). The agreement's purpose was to explore the possibility of implementing a property platform for digital payments. The telecommunication company also engaged my team and me in a new research project related to implementing another algorithm that was to become a new system: MBXI. Now, after many years, thinking back to that time, I can say that the payment system that I developed, implemented, and tested together with my collaborators (the engineers), Tiziana Landi and Alessandro Passerini, was one of the best things I have done in my career. Even though the MB09 payment system has never been adopted as a standard, you will find some interesting properties in the algorithm that might be useful to increase your ability as a cryptographer.

Now, if you follow me, I will guide you in discovering the first of the two algorithms, MB09. This is a method based on Fermat's Last Theorem and remodeled in a cryptographic mode. Then, we will analyze MBXI along with an interesting attack on RSA.

Introducing the MB09 algorithm

First, we will start with some considerations regarding the algorithm and reintroduce Fermat's Last Theorem. The first time I presented MB09, it was as an encryption algorithm, but effectively it is much more *a protocol for digital payments*. As I have already mentioned, while blockchain and cryptocurrency were not yet well known, I developed MB09 as an encryption/decryption algorithm to exchange a message between two actors. Many years later, I worked on the algorithm, taking it as the basis for a *fully homomorphic encryption* system and creating MB23, which was a *fully homomorphic algorithm*. Eventually, in 2020, it was turned into a new version, called **MBXX**, to overcome the consensus problem proposed by Satoshi Nakamoto.

Let's examine how the first version of MB09 worked. To do that, we'll recall Fermat's Last Theorem:

```
a^n + b^n = z^n
```

Here, the (n) exponents represent all the sets of positive integers.

This equation, as we already know, has been demonstrated to be never satisfied except for the n = 2 exponent (which is the well-known Pythagorean theorem), as follows:

```
3^2 + 4^2 = 5^2
```

Diving deeper to analyze this simple yet complex equation, we find that in modular mode, substituting the (n) exponents with a set of prime numbers, (p), the following is always verified:

```
a^p + b^p ≡ z^p (mod p)   (for all p>2)
```

This shows that we always have this equation verified for all primes >2. That means (at the opposite of its correspondent linear equation mode) for all of the (p) exponents, the sum of (a+b)^p is always verified equal to (z^p).

For example, consider the following:

```
3^17+ 5^17 ≡ 8 (mod 17)
```

So, we can re-write the preceding equation in the following form:

```
a^p + b^p ≡ (a+b)^p (mod p)
```

Alternatively, you can even write it in the following form:

```
a + b ≡ z (mod p)
```

That means, if we put a (p) prime number as an exponent and even the modulus with the same (p) value, then the preceding equation (in modular mode) returns to a linear mode so that (a+b) is always equal to its sum, (z).

The following equation is what Fermat's Last Theorem explains, and it determines the preceding linear function:

```
a^p = a (mod p)
```

For example, consider the following:

```
3^7 = 3 (mod 7)
```

You might not have noticed, but in the preceding function, we have used the (=) equal and (≡) not congruent notations. In this case, the meaning of (=) is very important because we intend to be equal in absolute values, so 3 is actually 3 and not only congruent.

I understand that all this could appear a little bit fuzzy, but try to follow me a little bit further ahead and the fog should disappear.

Fermat's Last Theorem states that $a^p = a$ (mod p) if a<p. So, what happens if we assume a>p?

For example, let's try, with $a = 5$ and $p = 3$:

```
5^3 ≡ 2 (mod 3)
```

As you can see, (a) is no more equal to (a) in its absolute value if a>p but results only congruent.

This intriguing property caused me to bear many considerations in mind regarding Fermat's Last Theorem, but far more (and that is important here) caused me to formulate a hypothesis:

"If I take a>>>p (much greater than p) and I keep [a] secret, then this will be a one-way function where it will be very difficult to return from the result, let's say from (A) to [a]"

Let's look at an example.

You can verify that $5^3 ≡ 2$ (mod 3), but also $8^3 ≡ 2$ (mod 3), and 11^3 .... 14^3, 17^3 are all congruent, and so on infinitely.

As you might have gathered, the curious thing is that starting from an initial value of 5 and adding 3 + 3 + 3... (essentially, adding 3 sequentially), we always obtain the same (2) value as the result in (mod 3), which is different in absolute value from (5) as Fermat's Last Theorem states p > a.

Finally, at the time, I wanted to demonstrate that is very difficult to come back from a public number (A) to a secret number [a] if p < a.

Let me explain this concept with an example taking a large [a]:

```
[a] = 96269369030336679694019965478670
```

That is, it would be easy to reduce this number to its corresponding modulo p=3.

In fact, we have the following:

```
96269369030336679694019965478670 ≡ 2 (mod 3)
```

Conversely, if I know the public number, (A) = 2, where the following is true:

```
A ≡ [a] ^ p (mod p)
```

Then, even if an attacker knows p = 3, it will be very difficult to attempt to obtain the [a] private key if a>>p:

```
[a] = 96269369030336679694019965478670
```

The simple reason for this is that p = 3 is our (Zp) ring, but [a], the private secret key, is outside of the (Z3) ring, and you don't know when to stop the iteration to find number [a]. Theoretically, it could be all the numbers from (p) to infinity and it doesn't matter how large [a] is because computationally is very easy to obtain (A) from [a] but not vice versa.

However, all that works but only in some instances.

Now that you know the basis on which MB09 leans, you will probably be curious to learn how it was implemented in its first version.

> **Note**
>
> I have represented, with a lowercase [a] letter, the hidden elements of the equation and, with the uppercase (A) letter, the known elements. However, don't be confused by this, and don't assume that A > a just because the letters are lower and uppercase. In fact, it's the opposite: in mathematical terms, A is much smaller than [a].

An explanation of MB09

As I mentioned earlier, the MB09 system is based on public/private key encryption principles. However, the scope here is *not* to send and receive a message between two (or more) actors. MB09 doesn't work correctly if used for this scope. However, the scheme of the algorithm works well for the scope to set up a protocol to manage and transmit digital cash in anonymity and secure the transactions. Eventually, one of the purposes of implementing such a system is that the network works as an *autonomous system*. I will explain this concept better, in the next section, when introducing MBXX, which is the evolution of MB09 in a decentralized environment. The version I'll present now, starting from a centralized system, will migrate to a decentralized and distributed system, as we will discover in MBXX.

In the next section, I will explain the basic concepts of a network governed by crypto algorithms, where Z (the centralized administrator) must verify whether all the transactions made by the users are correct and acceptable.

Z is our network's *centralized administrator*; you can think of it as a telecommunication provider or a bank. Alice and Bob are two actors in the network.

Bob is the sender, and Alice is the receiver. They can be considered virtual machines, servers, or computers. In our basic example, they are part of a network in which there are many users. The network's admin (Z) is a server that is linked to many users, as you can see in the following diagram of a centralized network:

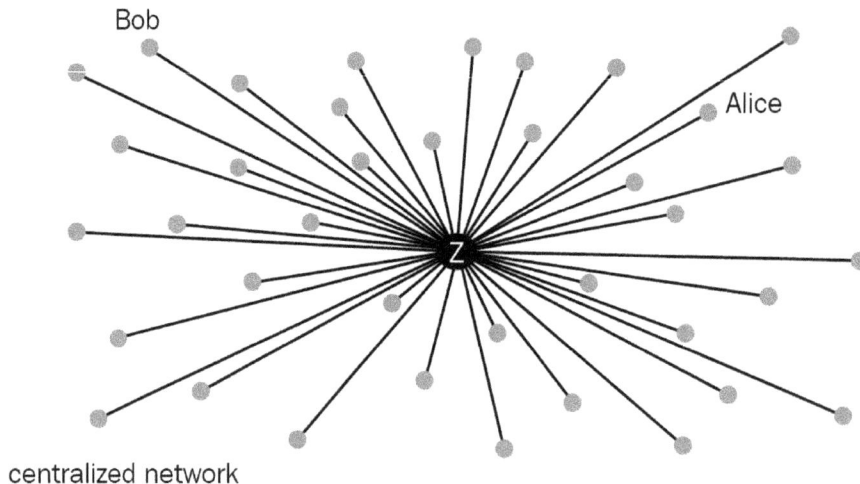

Figure 6.3 – A centralized network

Zeta has already published some parameters such as the (p) prime number and the public keys – (A) and (B) – of Alice and Bob and other participants.

We will proceed to look at the algorithm with only these two actors. Still, as I have already mentioned, this algorithm's strength is just the possibility of operating with many users who don't know each other, who don't trust each other, and who don't trust third parties. Indeed, Fermat's Last Theorem (which is extended in modular form), in this case, applied to MB09, is valid also for an infinite number of users of the network, as you can mathematically see here:

```
a^p + b^p + c^p + n^p ≡ z^p (mod p)
```

> **Important Note**
>
> I have illustrated the operations with a (+) sum; however, all of these operations could be performed by XOR, difference, or multiplication.

The [M] message has previously been encoded with ASCII code (as we learned in the first chapter), and you can think of [M] as an amount of digital money to transfer between (A) and (B).

Now, let's demonstrate whether Alice and Bob can transfer money to each other using the algorithm mathematically.

Each step of this algorithm after the *key initialization* consists of a recursive iteration of the following:

- Implementation of the public parameters of the users
- Operations on the parameters: the transmission of the messages (for example, a digital money transfer)
- Re-valuation of the new parameters

Key initialization:

[a] : This is the private secret key of Alice.

[b] : This is the private secret key of Bob.

This assumes that [a] and [b] are two random large primes.

Public parameters:

Alice and Bob calculate their public keys starting with their private keys:

```
A ≡ a (mod p)        [a>>p]
B ≡ b (mod p)        [b>>p]
C ≡ c (mod p)        [c>>p]
Z ≡ z (mod p)        [z>>p]
```

At this point, you can visualize the two f(z) and f(Z) equations. The first is in modular mode, and the second is transformed into linear mode:

```
[a]^p + [b]^p + [c]^p + … + [n]^p ≡ [z]^p (mod p)
A + B + C + … + N = Z
```

The correspondence between the first equation's elements and the second equation's elements is essential for the algorithm. As you can observe in the following diagram, each element of the first equation can be represented with an element of the second equation.

The peculiarity of this correspondence is that, while in the first equation, the operations have been performed using the users' private keys; in the second equation, the operations are performed with the public keys of the users. You can see that the arrows are only going in one direction, from [a] to (A), and it's difficult as we have seen to go back from (A) to [a] if [a>>p]:

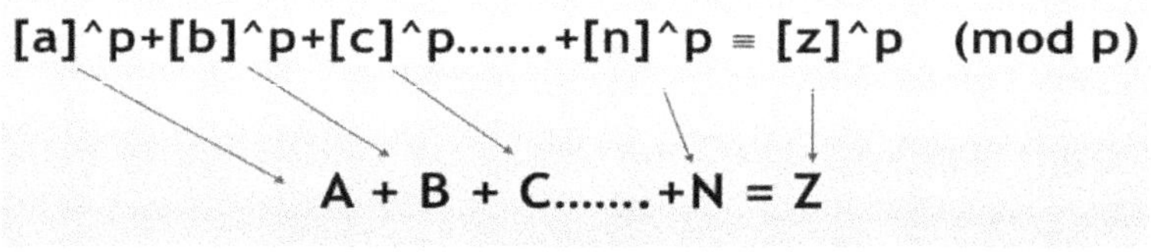

$$[a]^p+[b]^p+[c]^p\ldots\ldots+[n]^p \equiv [z]^p \quad (mod \; p)$$

$$A + B + C\ldots\ldots+N = Z$$

Figure 6.4 – The correspondence between the elements of two equations

Here, (Z) can be defined as the *isomorphic balance* between the operations performed with the parameters expressed clearly, instead of inside the square brackets, representing the secret parameters of the f[z] function.

At this point, MB09 (version 2009) went on with a standard cryptographic transmission based on a pair of keys to transmit a message between Alice and Bob.

> **Note**
>
> Remember that, here, [a] and [b] are Alice and Bob's secret values.
> Moreover, don't confuse the letters of [a] and (A), thinking that (A) is
> bigger than [a] just because the first is represented in capital letters.

First, we have to generate the secret transmission key, [K]. This is a shared transmission key that can transmit [M] between Alice and Bob. It can be generated with any public/private key algorithm, just like D-H algorithm.

Indeed, in the first version of MB09, I adopted the [K] key generated by D-H. However, as you might have gathered, it's possible to generate [K] with any other algorithm that uses the [a] and [b] secret keys of Alice and Bob as an input and gives a cryptogram, (c), as output.

> **Note**
>
> (c) is the cryptogram represented inside the round brackets. Here, I use (c)
> because we have only two actors. In other cases, I will use another notation –
> don't confuse it with an element of the network.

The mainframe of the algorithm with only two actors (but as I said, this is a multiple communication algorithm) is as follows:

```
[a]^p + [b]^p ≡ [z]^p (mod p)
(A) + (B) = (Z)
```

Alice encrypts [M] using her [Ka] secret key, where [Ka] has been generated by any public/private key algorithm, and obtains the cryptogram (c):

```
[M] * [Ka] ≡ (c) (mod p)
```

> **Important Note**
>
> Here, as I have already mentioned, (c) is the cryptogram, which is obtained
> through multiplication of the secret message [M] with Alice's private key
> [Ka]. In the original version of the algorithm, I used multiplication as the
> operation to generate (c). However, is possible to implement it with
> a different encryption method, such as Bitwise-XOR or scalar multiplication.

Bob decrypts, generating a [Kb] secret key such as the following:

```
c (INV) [Kb] = [M] (mod p)
```

For instance, if we assume the use of this algorithm in a digital cash environment, such as the transmission of digital cash in a payment system, [M] will be the real amount of digital cash transacted, while (Hm), its corresponding hash value, will represent the hash of [M]:

```
[a +/- M]^p + [b +/- M]^p ≡ [z +/- M]^p (mod p)
(A +/- Hm) + (B +/- Hm) = (Z)
```

The operations performed on the first equation (at the top) in encrypted gives the result [z] linked by an operation to the message [M] "in blind" that corresponds with the operations performed by the second equation's elements (bottom) and matches the result of the linear equation (Z). We will experiment with a similar protocol later in this chapter when we discuss the MBXX protocol.

Essentially, (Z) represents the *homomorphic balance* of the system, as the digital money transacted doesn't change the value of (Z). The scope to adopt such a homomorphic balance is that the administrator has, in each instance, a corresponding balance of the total amount of cash with the *isomorphic values* and can control the accuracy of the transactions even when the transacted amount, [M], is unknown.

Let's perform an exercise using a D-H key exchange:

- [a] and [b] are the secret parameters of Alice and Bob.

As we know, the public parameters in D-H are as follows:

- p: large prime number
- g: The generator of the (Zp) ring

The exchanging of [M] is performed as follows.

Step 1: Encryption:

Alice generates her (Ag) public key starting with her secret parameter, [a]:

```
Ag ≡ g^[a] (mod p)
```

Bob generates his (Bg) public key starting with his secret parameter, [b] :

Alice calculates the following:

```
Bg^[a] ≡ [Ka] (mod p)
```

Alice encrypts [M] with her [Ka] key:

```
[M] * [Ka] ≡ (c) (mod p)
```

Step 2: Decryption:

Alice sends (c, Ag) to Bob.

Alice sends the hash value of [M] = (Hm) to the administrator.

Bob decrypts (c), returning [M] :

```
( c) * [INVKb] ≡ [M] (mod p)
```

You can refer *Figure 6.5* for a complete scheme of the algorithm applied to the transmission of a secret message.

If we consider the private keys of Alice and Bob, [a] and [b] , as values of their accounts and [M] as the amount of digital cash transmitted, we have a third step that can be regarded as the homomorphic balance.

Step 3: The homomorphic balance:

Transposing the value of [M] both into the first equation and the second equation, the admin can verify that the transaction between (A) and (B) is correct:

```
[a +/- M]^p + [b +/- M]^p ≡ [z +/- M]^p (mod p)
(A +/- Hm) + (B +/- Hm) = (Z)
```

Pay attention to the following in the function of the encryption:

```
[M] * [K] ≡ (c) (mod p)
```

As I have said, I have used the multiplication between [M] and [K] to encrypt [M] . However, we can also implement the encryption with another operator, such as Bitwise XOR, between the message [M] and the key [K] . Note that, here, the algorithm of the transmission adopted to send [M] is not as important as the logical structure of the protocol for the implementation of multiple transmissions.

Indeed, the logical structure of this algorithm is not crafted for two users but multiple users.

In this scheme, it is supposed that the admin knows the amount of money originally held by the users. Let's look at an example:

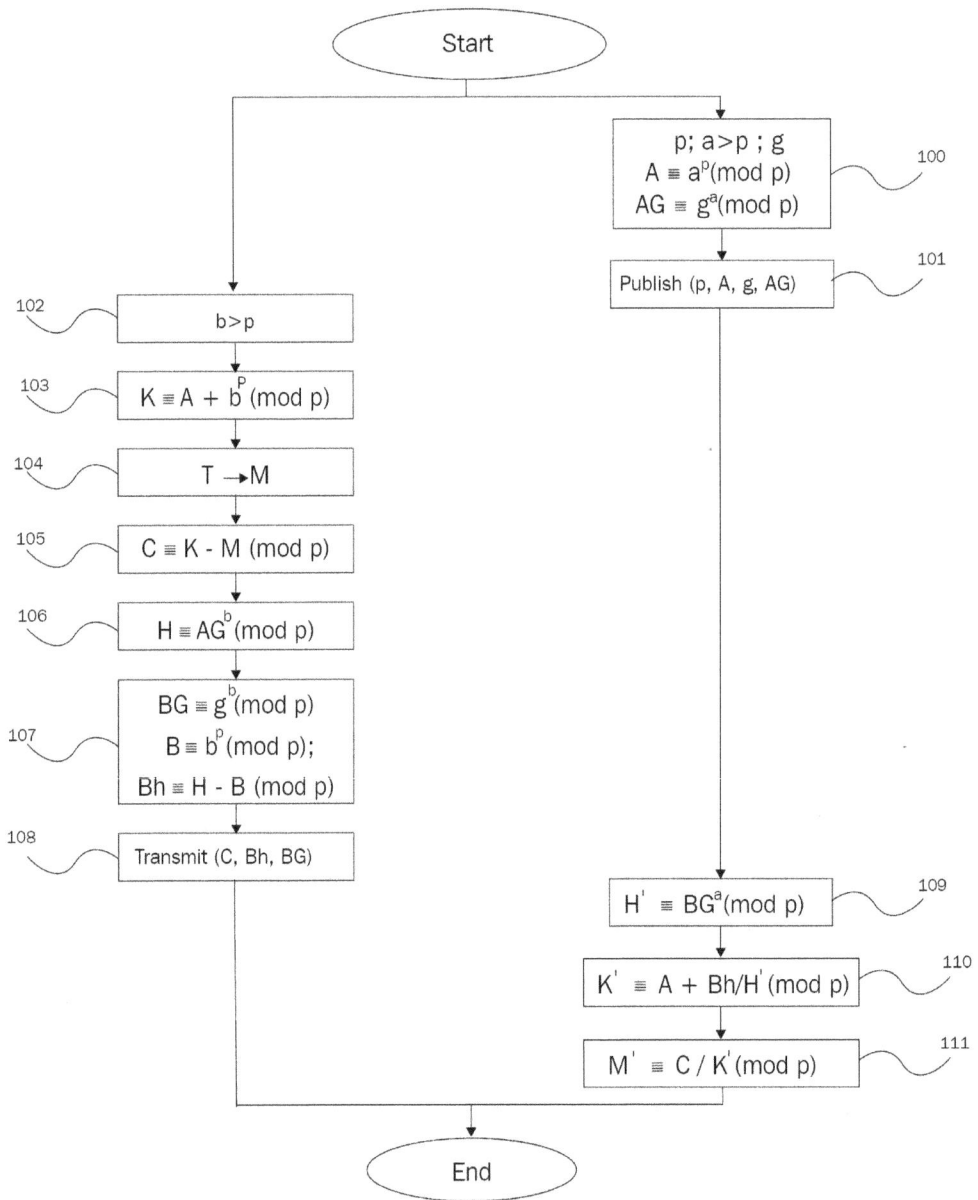

Start

$$p;\ a>p\ ;\ g$$
$$A \equiv a^p (\text{mod } p)$$
$$AG \equiv g^a (\text{mod } p)$$
100

Publish (p, A, g, AG)
101

102    $b>p$

103    $K \equiv A + b^p (\text{mod } p)$

104    $T \rightarrow M$

105    $C \equiv K - M (\text{mod } p)$

106    $H \equiv AG^b (\text{mod } p)$

107
$$BG \equiv g^b (\text{mod } p)$$
$$B \equiv b^p (\text{mod } p);$$
$$Bh \equiv H - B (\text{mod } p)$$

108   Transmit (C, Bh, BG)

109    $H' \equiv BG^a (\text{mod } p)$

110    $K' \equiv A + Bh/H' (\text{mod } p)$

111    $M' \equiv C / K' (\text{mod } p)$

End

Figure 6.5 – The original version of the patented MB09 published in the Patent Cooperation Treaty (PCT)

At the time, together with my colleagues, we crafted a **Proof of Concept** (**POC**) to demonstrate that MB09 could work well to exchange digital cash. We implemented a transmission of digital crypto-cash between cellphones. The protocol used was based on a modified version of MB09, which I have presented here. It had been added to the operative elements to engineer implementation.

In May 2010, the payment system based on MB09 was selected as a candidate for digital payments by a European telecommunication company. This company then acquired part of the rights on the patent of MB09 and entrusted our research laboratory with a research project to implement another private/public key algorithm, MBXI, which we will examine next.

Introducing the MBXI algorithm

When I started the MBXI project, it was 2010. I had already had the algorithm's scheme in my mind for some time before that, but there were some issues that I couldn't solve. While I was working on project MB09, to build a digital payment network, the telecommunication company that became my partner also needed to use a proprietary algorithm to send a message, [M]. In this scenario, that refers to the amount of digital money transmitted between two or more actors in the network.

So, I proposed a new public/private key algorithm. MBXI was patented in November 2011, and it's still valid now after ten years.

MBXI is a cryptographic process method that could be simultaneously considered both an asymmetric scheme and a symmetric scheme. We learned in *Chapter 2, Introduction to Symmetric Encryption*, how symmetric schemes work, and in *Chapter 3, Asymmetric Encryption*, we learned how asymmetric schemes work. This algorithm bases its strength on the discrete logarithm problem, which has already been explored across different algorithms in this book (for example, D-H, ElGamal, zero-knowledge protocols, and more). As mentioned, the discrete logarithm problem is still a very hard problem to solve; we have learned that if modular equations (such as the p modulo) are implemented using a big prime number, the solution of exponential modular equations is very burdensome or almost impossible.

This algorithm derives its robustness from the application of modular exponential equations, which are injected (one-way) functions for defining exponents of the encryption (and decryption) equations. As you can see in the flowchart (*Figure 6.6*), the [T] message is decomposed into [M] sub-messages divided into blocks.

In MBXI, the only way to determine the private key is to solve the discrete logarithm differently with other asymmetric encryption algorithms, such as RSA, that suffer from both the problems of factorization and discrete logarithm (as explained in *Chapter 3, Asymmetric Encryption*), along with ElGamal.

In fact, the best way to define this algorithm is as neither asymmetric nor symmetric but a public/private key algorithm; I will explain the reasons for this later in this chapter.

Let's dive deeper to explore the scheme of MBXI.

Alice and Bob are always our two actors. In this scenario, let's consider that Bob wants to send an [M] secret message.

The first step of the algorithm is to generate the private and public keys given by the common public parameters published on the network:

- p: This is a large prime number.
- g: This is a generator of the ring (Zp).

Step 1: **Key generation**:

- [a]: This is the private secret key of Alice (a large number chosen inside p-1).
- [b]: This is the private secret key of Bob (a large number chosen inside p-1).
- Alice's public key: Ka ≡ g^a (mod p).
- Bob's public key: Kb ≡ g^b (mod p).

Step 2: **Encryption**:

The encryption in MBXI is done by an inverse modular equation expressed by solving the following function:

```
{[Ka^b+eB] (mod p)}*x ≡ 1 (mod p-1)
```

In this inverse equation, Bob takes the public key of Alice, (Ka), elevates it to his private key, [b], and summed to (eB), where (eB) is a parameter selected from the group of integers so that the co-primeness between the following is verified:

```
[Ka^b+eB] (mod p) and (p-1)
```

Then, after solving the inverse modular equation for [x], Bob calculates the cryptogram, (C), that is given by the function:

```
C ≡ M^x (mod p)
```

Bob sends the triple (C, eB, Kb) function transmission to Alice.

Step 3: Decryption:

Alice can now decrypt (C) using Bob's public key, (Kb), and her private key, [a] along with the parameter (eB). This parameter is transmitted by Bob, whose function will be better described later during the analysis of the algorithm.

The decryption consists of solving the following function, which produces [y] as Alice's private decryption key:

```
y ≡ { [Kb^a + eB] (mod p) }
```

Finally, the [M] message is returned to Alice, elevating (C) to [y]:

```
M ≡ C^y (mod p)
```

Where, as mentioned, [y] is Alice's private decryption key, represented by the solution of the preceding inverse modular equation.

The flowchart in the following diagram will help you to gain a better understanding of the entire process of the encryption/decryption of MBXI:

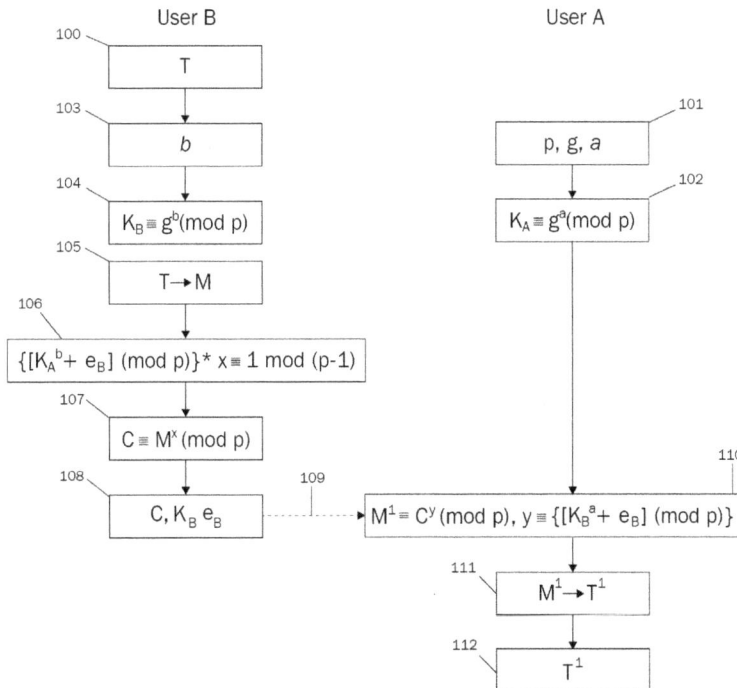

Figure 6.6 – The encryption/decryption process in MBXI

Now that we have examined how MBXI works, let's look at an example with numbers to understand it better.

A numerical example of MBXI

The following numerical example represents a cryptographic communication established between a sender (Bob) and a receiver (Alice). Of course, like all the other examples, we use relatively small numbers. Bear in mind that, in a real application, the private keys are, at the very least, of the order of 3,000 bits (that is, thousands of digits), such as the (p) prime. It is defined as a prime number:

- p = 7919.

- g = 7 is the generator.

Alice selects her private key, [a]:

```
a = 123456
```

Bob selects his private key, [b]:

```
b = 543210
```

> **Important Note**
>
> As you have noticed, I have selected a sequence of numbers for a = 123456 and b = 543210. This is just an example and you should never use a sequence of numbers when you select a password in the real world because this is the first combination of numbers tried by an attacker.

According to the first step of the algorithm (key generation), Alice computes her public key as follows:

```
Ka ≡ 7^123456 (mod 7919) = 7036
```

Bob does the same, who computes his public key as follows:

```
Kb ≡ 7^543210 (mod 7919) = 4997
```

Bob's encryption is performed in accordance with the following equation:

```
{ [7036^543210 + 1 ] (mod 7919)} * x ≡ 1 (mod 7919-1)
```

Here, (eB) = 1 is the smallest integer that verifies the equation.

Therefore, Bob determines the following private encryption key [x]:

```
x = 3009
```

With this parameter, Bob calculates the cryptogram, (C):

```
C ≡ 88^3009 (mod 7919) = 2760
```

Bob sends the triple (C, Kb, eB) = (2760, 4997, 1) transmission to Alice.

Once Alice has received the triple (C, Kb, eB) transmission, she is able to decrypt cryptogram (C) and determine the message block of [M]:

```
M = 2760^{ [4997^123456 +1] (mod 7919)} (mod 7919) = 88
```

Indeed, M = 88 matches the original message block, [M], transmitted by Bob.

As you can observe, in the preceding example, I have demonstrated with real numbers that the algorithm works. Now, I will offer some notes on it to point out some functions and show similarities with RSA.

Notes on the MBXI algorithm and the prelude to an attack on RSA

First of all, I want to clarify the meaning of the (eB) parameter. This is clearly transmitted between the sender and the receiver. However, in different versions, it can be randomly processed.

For example, (eB) is randomly selected from the range of (1, 10). Let's assume that (eB) is 6. Therefore, the equation of encryption will be as follows:

```
{ [7036^543210 + eB] (mod 7919)} * x ≡ 1 (mod 7919-1)
```

Note that for (eB) = 6, the equation is *not* verified!

Indeed, if we try to put eB = 6 in the preceding function instead of eB = 1, using mathematica Reduce function, this is the result:

```
Reduce[4960*x == 1, x, Modulus -> 7919 - 1]
False
```

In other words, the inverse number for 4960 (mod 7919) doesn't exist. In addition to this, the number that is randomly selected from the given range is increased by one, and the check process is repeated using (eB) = 7, which satisfies the co-primeness condition whereby x = 5575.

So, we will have the following encryption process:

```
C ≡ 88^5575 (mod 7919) = 2195
```

In this case, Bob's triple transmission to Alice will be (C, Kb, eB) = (2195, 4997, 7).

So, the decryption will be as follows:

```
M ≡ 2195^{[4497^123456 + 7] (mod 7919)} (mod 7919)
```

From this equation, M = 88 matches the original message block of [M] that was transmitted by Bob.

Another important consideration relates more to the implementation of MBXI and, in particular, to several problems related to the length of the [M] message.

The [M = 1] message gives the ciphertext of (C = 1), no matter what the encryption key.

Similar properties are true of RSA in its simplest form. Most RSA used in practice includes message padding, which eliminates these properties, and the same could be done with MBXI. For some very specialized uses, the preceding properties could be helpful – the technical term is *homomorphic encryption*. We will analyze the partial homomorphic property of RSA in *Chapter 9, Crypto Search Engine*.

As discussed in *Chapter 5, Introduction to Zero-Knowledge Protocols, padding* is necessary to remove the (generally) undesirable properties. This means that a ciphertext string will be longer than the underlying message string.

You might be wondering whether MBXI is an authentic asymmetric encryption algorithm. My answer to this question is no; MBXI is a public/private key encryption algorithm but not a pure asymmetric encryption algorithm such as RSA. Even if this peculiarity of MBXI has some interesting applications, for example, to share a key, it was absolutely not the scope of MBXI to substitute the power of a symmetric algorithm such as AES.

To demonstrate the scheme of the shared key, I will reformulate the equation that determines the encryption step:

```
C ≡ M^x (mod p)
```

Suddenly, we notice a similitude with RSA's encryption structure:

```
C ≡ M^e (mod N)
```

It's just a matter of changing the `[x]` parameter with `(e)`, and it turns out that the two encryption schemes are similar.

In RSA, remember that `(e)` is a public parameter, and `(N)` is the public key of the receiver, given by `N = [p] * [q]`.

In contrast, in MBXI, `[x]` is a secret encryption key (more similar to D-H), and `(p)` is a big public prime number.

Anyway, these comparisons will be useful to explain a couple of interesting attacks on RSA, which has inspired me due to the similarity of the RSA and MBXI encryption schemes.

Unconventional attacks on RSA

The notion of an asymmetric backdoor was introduced by Adam Young and Moti Yung in their *Proceedings of Advances in Cryptology* paper. An asymmetric backdoor can only be used by the attacker who plants it, even if the full implementation of the backdoor becomes public (for example, via publishing, being discovered and disclosed by reverse engineering, and more). This class of attacks has been termed kleptography; they can be carried out on software, hardware (for example, smartcards), or a combination of the two. The theory of asymmetric backdoors is now part of a larger field called **cryptovirology**. Notably, NSA inserted a paragraph on kleptographic backdoor into the Dual EC DRBG standard.

There exists an experimental asymmetric backdoor in RSA key generation. This OpenSSL RSA backdoor, designed by Young and Yung, utilizes a twisted pair of elliptic curves and has been made available. In this section, we will examine something that is a little bit different because it involves the injection of a parameter inside the `Modulo N` of the RSA encryption function. Likewise, it's possible to modify a few rows in the code of the OpenSSL library, and as a result, we will achieve the creation of a backdoor in RSA.

A backdoor is something that allows an attacker to spy on the communications between two or more people, and in cryptography, a backdoor allows the attacker to decrypt an unbreakable cipher by adopting a mathematical or physical implementation method.

In both cases, creating a backdoor needs a certain allocation of resources in terms of skills and preparation to inject the malware. Eventually, the result is to allow a third party (Eve) to spy on or modify the encrypted message. However, there is also another possibility to create an even more malicious backdoor than others: the sender himself (Bob) might have an interest in injecting a backdoor inside his own encryption.

> **Important Note**
>
> Implementing a backdoor for malicious purposes is a crime. So, I am warning
> you to consider this section as merely a theoretical way to understand
> that there is the possibility of implementing a backdoor inside RSA and
> contrasting it.

It's time to return to a question posed in *Chapter 3, Asymmetric Encryption*: is this method of *Self-Reverse Encryption* an unreasonable model, and not representative of practical situations, or does it, instead, have practical uses? In other words, why or in which cases could it be reasonable to use a method like this to attack a network?

You can think of Bob as the administrator of a telecommunications company, a social network, or any cloud provider who is willing to spy on communications between users. But even in the cryptocurrency space, there should be interesting cases if Bob, the sender (in the case of a cryptocurrency payment), is able to perform a fake signature on the amount transmitted. He probably will be able to perform *double-spending* of this digital money. Finally, the self-reverse encryption method could be used (in this case for a good purpose) to retrieve the encrypted files of a ransomware attack.

As Simon Sight wrote in his 1998 book, *The Code Book*, a report on the *Wayne Madsen Report* blog revealed that a Swiss cryptographic company, Crypto AG, had built backdoors into some of its products and had provided the US government with details of how to exploit these backdoors. As a result, America was able to read the communications of several countries. In 1991, the assassins who killed the exiled former Iranian prime minister were caught; this was thanks to the interception and backdoor decipherment of Iranian messages encrypted using Crypto AG equipment.

Now, let's explore how to create a logical backdoor using some of the functions we have learned in this book.

We'll start from one of the main paradigms and constraints in the RSA encryption stage. This paradigm states that once the cryptogram has been sent by Bob, the sender, it cannot be re-decrypted by the sender themselves. That happens because only the receiver (Alice), the owner of the (p,q) private keys, can perform (N). Consequently, she is the only one who can decrypt the (c) cryptogram and receive the message $[M]$.

Let's refresh our knowledge from *Chapter 3, Asymmetric Encryption*. In the following diagram, you can see the RSA algorithm encryption stage:

RSA: scheme of Encryption/Decryption

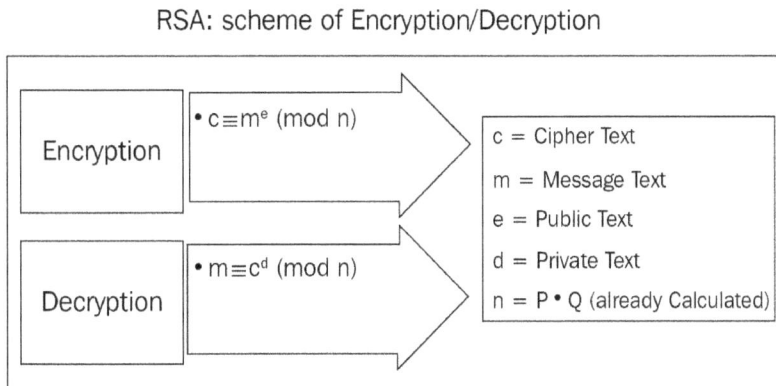

Encryption	$\bullet\ c \equiv m^e$ (mod n)	c = Cipher Text
		m = Message Text
		e = Public Text
Decryption	$\bullet\ m \equiv c^d$ (mod n)	d = Private Text
		n = P $\bullet$ Q (already Calculated)

Figure 6.7 – The RSA encryption stage

In this stage, we have the following:

- [M] : is the secret message
- (N) = p*q (Alice's public key)
- e: is a public parameter (given)

I would remark that the encryption is performed by the sender (Bob) using Alice's public key, which we will now rename (Na) to clarify that it is Alice's public key:

```
[M]^e ≡ c (Mod Na)
```

Bob, the sender, knows the secret message, [M], but technically, he is *not* able to *re-decrypt* mathematically the (c) cryptogram once he delivers the (c) cryptogram. That's because it's assumed that only Alice knows her [da] private key.

Key generation:

Alice's public key, (Na), is given as follows:

```
Na = p*q
```

[da] is given as follows:

```
[da] * e ≡ 1 (mod [p-1]*[q-1])
```

Recall from *Chapter 3*, *Asymmetric Encryption*, that this last inverse multiplication is the core of RSA's algorithm because it allows you to retrieve the secret message, [M], hidden inside the cryptogram, (c), through the following operation of decryption:

```
M ≡ c^da (mod Na)
```

Now it should be clear that one of the main paradigms of the RSA algorithm is that the cryptogram, (c), once it has been delivered by the sender, cannot be decrypted by anyone who isn't the holder of the [da] private key.

However, I can demonstrate that this statement is not applicable in some particular circumstances, when a backdoor is injected.

Now, let's move on to analyze how to implement this pseudo-attack, which is able to generate self-reverse decryption (by Bob) of the cryptogram in RSA.

Remember again, here we assume that Bob himself (the sender) is the author of the attack. Then, as you will see, it's easy to demonstrate that Eve, an external attacker, can also replace Bob.

The public and private keys:

- (Na) : This is the public key of Alice (the receiver).
- (Nb) : This is the public key of Bob (the sender).
- (e) : This is a public parameter (given by the system).
- [da] : This is Alice's private key.
- [db] : This is Bob's private key.

Bob performs the "attack" by changing the parameters by himself.

Step 1: Encryption:

Bob chooses a prime number, [Pb] > [M], such that he is able to perform a modified encryption of (c1):

```
[M]^e ≡ c1 (mod Na* Pb)
```

He sends (c1) to Alice.

> **Note**
>
> The [Pb] parameter is the backdoor that allows Bob to return the [M] message.

Step 2: Bob's decryption:

At this point, Bob calculates the [x] parameter, which is given by the multiplicative inverse of the function:

```
e*x ≡ 1 (mod Pb-1)
```

The result of this inverse multiplication gives back [x], a special private key of Bob, which gives Bob access to the decryption of (c1). This is the "magic function" used by Bob to retrieve the [M] message.

Bob can now decrypt (c1) using [x] as his "private key" through the following function:

```
(c1)^x ≡ [M] (mod Pb)
```

Bob can now mathematically retrieve the [M] message from the (c1) cyphertext. That goes against what is commonly stated by RSA: it is impossible for the sender to decrypt the cryptogram once it has been performed because the sender doesn't hold the receiver's private key.

Someone could counter that by saying that we have changed the RSA encryption, so we are dealing with a different algorithm and no longer with RSA. That is true to some extent but let's see what happens when Alice decrypts (c1).

Step 3: Alice's decryption:

Here comes the interesting part of this method: once Alice receives the fake cryptogram, (c1), she is able to decrypt (c1) with her private key, [da], and retrieves the secret message, [M], keeping the RSA decryption stage unaltered:

```
(c1)^da ≡ [M] (mod Na)
```

Alice gets the [M] message by performing the RSA decryption, unaware of the fake (c1) cryptogram received.

As you can see in the following diagram, Bob can perform his self-reverse decryption of (c1) "modified cryptogram," while Alice uses her private key to decrypt the fake (c1) cryptogram:

Bob's Self-Reverse Decryption in RSA

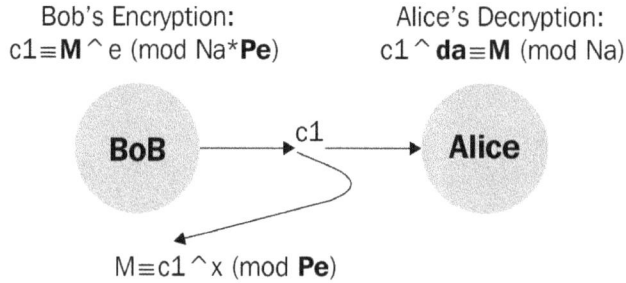

Bob's Encryption:
$c1 \equiv M \char`\^ e$ (mod Na*Pe)

Alice's Decryption:
$c1 \char`\^ da \equiv M$ (mod Na)

BoB   c1   **Alice**

$M \equiv c1 \char`\^ x$ (mod Pe)

Figure 6.8 – A schema of Bob's self-reverse attack

It's easy to demonstrate that if this attack is performed by Eve (an external attacker) who is able to inject her [Pe] parameter (backdoor) in Bob's encryption stage, she can spy on the [M] message exchanged between Bob and Alice.

In the following diagram, you can see the schema of the attack (spying) taken from Eve:

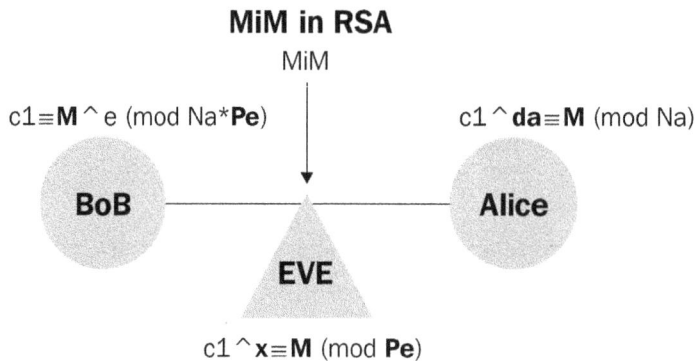

MiM in RSA

MiM

$c1 \equiv M \char`\^ e$ (mod Na*Pe)

$c1 \char`\^ da \equiv M$ (mod Na)

BoB   **Alice**

EVE

$c1 \char`\^ x \equiv M$ (mod Pe)

Figure 6.9 – A schema of Eve's spying attack (MiM)

With this scheme, not only can Eve retrieve the [M] message, but she can also modify it by sending a fake new message, [M1], to Alice, deceiving Bob and Alice, and then recreating a new RSA encryption (original) after having read the message sent by Bob.

To make that possible, Eve has to face off with a problem: the counting of bits in the OpenSSL library. That is because the (fake) encryption stage generates an overflow of bits in (c1) that is much larger than the original (c) cryptogram. In mathematical terms, c1 >> c. This could be a problem if there is a preselected *limit* of bits given by the parameters of the algorithm. This limit could be related to the key length or other parameters in the algorithm, for example, the N modulo. In other words, if N = 2000 bits, then (c1) cannot be bigger than 2000 bits; otherwise, (c1) is not accepted.

I found a way to avoid this problem by adopting a compression algorithm of my invention, able to generate an output cryptogram in the (cx) < (Na) encryption stage. In this way, the cryptogram generated by the fake (c1) encryption stage is reduced to (cx) and accepted as a *good* parameter by the OpenSSL library, overcoming the problem of the bit-length limit. This compression algorithm can also be used for other scopes: for example to compress and recompose a cryptogram in transmission to save bandwidth.

As you can see in the following diagram, Eve can spy on the [M] message transmitted by Bob and also modify it with [M1], sending to Alice a new cryptogram, (ce), re-generated with an original RSA encryption stage:

Compression Algorithm for RSA

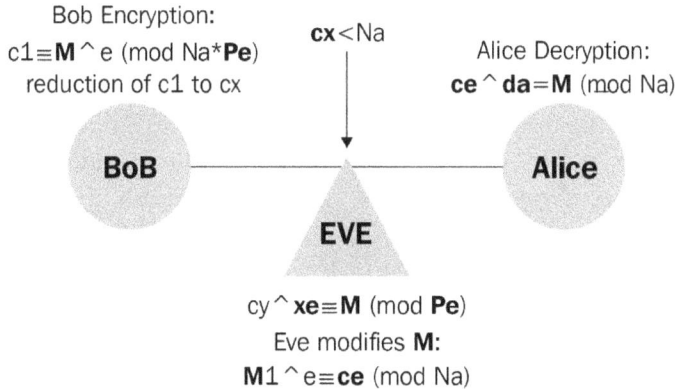

Bob Encryption:
$c1 \equiv M \char`^ e \pmod{Na*Pe}$
reduction of c1 to cx

$cx < Na$

Alice Decryption:
$ce \char`^ da = M \pmod{Na}$

BoB ———————— Alice

EVE

$cy \char`^ xe \equiv M \pmod{Pe}$
Eve modifies M:
$M1 \char`^ e \equiv ce \pmod{Na}$

Figure 6.10 – Eve attacking through the compression algorithm

Now we should be able to answer another question: if a digital signature is requested from Alice on the [M] message, is Eve able to perform this attack?

In other words, is there some way to counter these attacks?

I will answer this question by saying that there are multiple ways to avoid these attacks depending on who performs the attack. One of the remediations to avoid these attacks is related to digital signatures (we looked at digital signatures in *Chapter 4*, *Introducing Hash Functions and Digital Signatures*). However, in this case, a special kind of digital signature is required to counter-fight these attacks.

Indeed, if Alice requests Bob to digitally sign the (c) cryptogram with his [db] private key, then Alice will be sure that no one has spied on or modified the [M] message and the (c) cryptogram was truly signed by Bob and performed with the original RSA encryption.

In fact, Bob signs the (c) cryptogram using his private key [db]:

```
c^[db] ≡ Sb (mod Nb)
```

Bob sends (Sb) to Alice, and Alice can verify whether the following condition has been met:

```
Sb^e ≡ cv (mod Nb )
```

With this operation, Alice finds the cryptogram: cv (cryptogram verification).

Then, if Alice finds out that the message elevated to the public parameter (e) corresponds to the cryptogram received, so cv = c, then she can be sure enough that no backdoor will be injected:

```
[M]^e ≡ c (mod Nb )
If: cv = c then the message is accepted by Alice.
```

Indeed, performing this operation, Alice compares the cryptogram received with the cryptogram obtained from decrypting the (Sb) signature. If it is the same, then probably no backdoor was injected by a third party. If Alice obtains cryptogram (c1) or (cx), or another cryptogram instead of (c), she understands something is wrong.

Note

Here, the signature is performed on the (c) cryptogram, not on the [M] message as is normally done.

In the next section, we will explore the digital signatures on the MBXI algorithm, and we will learn that there are different ways in which to digitally sign this algorithm.

Digital signatures on MBXI

Returning to MBXI, we notice that [x], the reformulated encryption key, is able to perform the encryption:

```
C ≡ Mˣ (mod p)
```

[x] results in the inverse of [y], the decryption key, in the following function:

```
C^y ≡ M (mod p)
```

In mathematical language, the encryption equation is as follows:

```
{[Ka^b+eB] (mod p) * x ≡ 1 (mod p-1)
```

This results in the inverse of the decryption equation, [y]:

```
y ≡ {[Kb^a+eB] (mod p)}
```

Let's perform a test with numbers to understand it better:

- x = 3009
- y = 4955

If we input x = 3009 in the inverse function (mod p-1), we can find the result [y] using Mathematica:

```
Reduce [3009*x == 1, y, Modulus -> p - 1]
y == 4955
```

That means, if Bob sends a message using MBXI, he will share a [secret key] type with Alice.

Another problem arises: how is it possible to avoid a MiM attack in a symmetric algorithm?

As you can see, MBXI has more characteristics of an asymmetric algorithm than a symmetric algorithm, so let's go and analyze the algorithms of the digital signature for MBXI.

As we have learned, MBXI could be defined in the same way as D-H, a shared key algorithm. However, in MBXI, in contrast to D-H, it is possible to perform a digital signature; in fact, we will discover that with MBXI there are different modes to digitally sign the message.

For more clarity, I want to recall the scope and functions of the digital signature.

The digital signatures should satisfy the following conditions:

- The receiver can verify the sender's identity (authenticity).
- The sender cannot disclaim the transmission of a message (no repudiation).
- The receiver cannot invent or modify a document signed by someone else (integrity).

Moreover, after you have learned about my backdoors creation, let me add another condition that the digital signatures should avoid:

- Anyone not authorized can't spy into the document of someone else (no spying).

This last condition will be one of the principal objects of our study. So, a typical scheme of our digital signatures consist of three steps:

1. An algorithm for key generation, (G), that produces a couple of keys, { (Pk) and [Sk] }, where (Pk) is the public key for the verification of the signature and [Sk] is the secret key owned by the signatory used to sign the [M] message.
2. An algorithm for the signature, (S), that takes the message, [M], as the input (usually the hash of the message) and a secret key, [Sk], to produce a signature, (s).
3. A final algorithm for verification, (V), takes the [M] message as input (usually, this is the hash of the message), the public key, (Pk), and a signature, (s), and based on that, it accepts or rejects the signature.

Given all these conditions and constraints, our digital signatures could be of two types:

- **A direct signature**: It is a scheme to retrieve the [M] message directly from the signature function (S). It is commonly used in RSA to perform the digital signature (S) on the message, as we have seen in *Chapter 4, Introducing Hash Functions and Digital Signatures*:

    ```
    [M]^d ≡ S (mod N)
    ```

And its correspondent verification stage is: $S\char`^e \equiv M \ (mod \ N)$

- **A signature with an appendix**: It doesn't need the original message to verify its validity (an example is ElGamal's signature), in which the [M] message doesn't need to be signed directly to verify the signature.

Now that we have fixed the concepts of digital signature methods, we can dive deeper to analyze any single case applied to MBXI.

A direct signature method in MBXI

The direct signature is a method that directly involves the message [M] signature. However, as we have seen in *Chapter 4, Introducing Hash Functions and Digital Signatures*, in some cases, it is much better to apply the hash of the message H(m); otherwise, it could be possible to recover the original message [M] from (S).

Let's see how this method works in MBXI. From the MBXI algorithm, we have the following parameters:

- Alice's private key: [a] (SKa)
- Alice's public key: (KA) (PKa)
- Bob's private key: [b] (SKb)
- Bob's public key: (KB) (PKb)

Bob's public key is given by the following function:

```
KB ≡ gᵇ (mod p)
```

Assuming Bob sends a message, [M], he will choose a parameter, [eb], known by Alice and Bob only. That's because as we said, the function of encryption is verified only under determinate values of [eb]:

```
(E): {[K_A^b + e_B] (mod p)} * x ≡ 1 mod (p-1)
```

Bob and Alice convene to determine [eb] autonomously, in the sense that [eb] is generated step by step, through a process that I have called *joint iteration*. This process consists of a progressive iteration of [eb] until the functions of encryption results are verified.

So, we can reformulate (as mentioned earlier) the preceding encryption equation into the following mode:

```
C ≡ Mˣ (mod p)
```

Here, [x] is the result of the preceding functions of encryption.

If [M] is the message sent in the session and H(m) is its corresponding hash value, Bob can sign H(m) in the same way he performed the encryption:

$$S \equiv H(m)^x \pmod{p}$$

Here, (S) is the digital signature.

When Alice receives the (C) cryptogram, she will also receive the (S) signature calculated with the hash of the corresponding [M] message.

Alice can decrypt the (S) signature with the same equation that was used to decrypt [M] but with a different [eb]. As we have discovered, Bob has iterated [eB], bringing its value into the next step of verification of his encryption function:

$$H(m) \equiv S^y \pmod{p}$$

Here, [y] is given by the following decryption function:

$$y = \{ [K_B^a + e_B] \pmod{p} \}$$

As you can see, Alice verifies the (S) signature with the public parameter of Bob, (KB), combined with her secret key [a].

Nobody other than Bob can be the sender. An attacker could only replace Bob's private key [b] in order to perform the following signature equation (S):

$$(S): \{ [K_A^b + e_B] \pmod{p} \} * x \equiv 1 \bmod (p-1)$$

However, that means an attacker can perform the Discrete Logarithm that returns from (KB) Bob's private key [b]. But as we have seen with our knowledge, that is a very hard problem now.

Indeed, remember that [x] is the result of the preceding function (S) and it needs Bob's private key [b] to be performed.

But let's examine another method of digital signature to apply to MBXI.

The appendix signature method with MBXI

An alternative method to signing the message with MBXI is called a **signature with an appendix**.

The signature with appendix is a method that we have already found in ElGamal (*Chapter 4*, *Introducing Hash Functions and Digital Signatures*). The MBXI algorithm arrives at determinate equality of results, which proves the truth of the signature.

From the MBXI algorithm, we, again, have the following parameters:

- Bob's private key: [b] (SKb)
- Bob's public key: (KB) (PKb)

These are given by the following function:

$$KB \equiv g^b \ (mod \ p)$$

The scope of the signature algorithm, (S), is to produce proof, (s), given by Bob such that Alice (the receiver) can verify, (V), its authenticity.

To make this consistent, Bob creates a hash from the secret [M] message such that it will not be possible for anyone who doesn't know the original message to recover [M] from H(m).

Hence, Bob wants to sign H(m) to demonstrate effectively he is who he claims to be.

Step 1: **Key generation (G)**:

Bob chooses a random number, [k], in the (Zp) ring. Then, he calculates (r):

$$r \equiv g^k \ (mod \ p)$$

Step 2: **Digital signature (s)**:

Bob computes the digital signature (s) based on the hash of the H(m) message combined with the random key [k] and his private key [b] as the following:

$$s \equiv H(m) * [k+b] \ (mod \ p-1)$$

Bob sends the triple (H(m), s, r) transmission to Alice.

Step 3: **Verification of the signature (V):**

Alice verifies the following two functions (V) and (V1):

```
(V)  g^s ≡ V (mod p)
(V1) r^H_(m)  *  Kb^H_(m)  ≡  V₁ (mod p)
```

If $V = V_1$, Alice accepts Bob's signature.

> **Important Note**
>
> The random key [k] has to be kept secret and changed at each digital signature session.

Now that we have examined the different methods of digital signatures in MBXI, let's look at a mathematical demonstration of the digital signature in MBXI.

A mathematical demonstration of the MBXI digital signature algorithm

In this section, I will demonstrate how the digital signatures in MBXI work mathematically.

Indeed, for the properties of power elevation, elevating the (g) generator to the equation in *step 2* covered earlier, we have the following:

```
g^s ≡ g^[k+b]^H(m) ≡ (g^k)^H(m)  *  (g^b)^H(m)  (mod p)
```

From this, we can substitute gk = r and gb = Kb and obtain the following:

```
g^s ≡ r^H(m)  *  Kb^H(m)  (mod p)
```

That is what we wanted to demonstrate.

Now, let's observe what happens with numbers to better understand how these digital signatures work.

A numerical example of the signature methods in MBXI:

If we take, as input, the same parameters of the previous example of encryption, we have the following:

- p = 7919
- g = 7
- eB = 1

This is Alice's private key, [a]:

- a = 123456

This is Bob's private key, [b]:

- b = 543210

The public key of Alice is as follows:

```
Ka ≡ 7^123456 (mod 7919) = 7036
```

And the public key of Bob is as follows:

```
Kb ≡ 7^543210 (mod 7919) = 4997
```

Recall that the [M] message encrypted by Bob was as follows:

```
M = 88
```

First, let's explore the direct signature (S).

Step1: Bob calculates the hash of the [M] = 88 message; suppose that the following is the result:

```
Hash(88) = 1305186650
```

Using our well known Mathematica Reduce function, we gain [x], the secret key of encryption:

```
Reduce[(Mod[KA^b, p] + eB)*x == 1, x, Modulus -> p - 1]
x = 3009
```

Using H(m) combined with [x], Bob calculates the digital signature (S):

```
S ≡ 1305186650^[3009]  (mod 7919)  = 7734
```

Bob sends (H(m), S) = (1305186650, 7734) to Alice.

Step 2: Next, Alice can verify whether the (S) signature effectively corresponds to Bob.

Elevating the signature to [y], the result must correspond with H(m) (mod p):

```
S^y ≡ H(m)  (mod p)
```

As we have seen, [y] is the decryption function given by the public key of Bob, (Kb), elevated to Alice's secret key, [a], added to the [eB] parameter:

```
y ≡ { [K_B^a + e_B]  (mod p) }
Substituting in the equation we have:
y ≡ { [4997^123456 + 1]  (mod 7919) }
Y = 4955
```

Alice takes (S) and elevates it to [y]. The result should be the corresponding value of H(m) (mod p):

```
V ≡ 7734^4955 = 827 (mod 7919)
```

That corresponds to the following:

```
H(m)  (mod p)
V' ≡ 1305186650 (mod 7919)  = 827
```

If V = V', Alice accepts the signature.

Indeed, Alice can verify that:

```
V = 827 = V'
```

As I have explained in this section, there is another way to sign and verify a signature with MBXI.

Let's view an example by numbers of the signature with the appendix.

All the parameters (modulo, private and public keys, generator, and eB) remain the same as before.

Bob picks up a random number, [k]:

```
k = 1529
```

Bob calculates (r):

```
r ≡ g^k (mod p)
r ≡ 7^1529 (mod 7919) = 4551
```

Now Bob is able to perform the (s) appendix signature:

```
H(m) = 827 (mod p)
s ≡ H(m) *[k+b] (mod p-1)
s ≡ 827 * [1529 + 543210] (mod 7919-1) = 4543
```

Bob sends (H(m), s) = (827, 4543) to Alice.

Note that, here, to work with low numbers, I have calculated the modulo (p) of H(m).

Alice verifies the following:

- V ≡ r^H(m) * Kb^ H(m) (mod p)

- V ≡ 4551^827 * 4997^827 (mod p) = 7147

- V' ≡ g^s (mod p)

- V' ≡ 7^4543 (mod 7919) = 7147

- V = V'

So, Alice accepts the (s) signature.

As we have completed the discussion about digital signatures in the MBXI algorithm, I want to present the evolution of the two algorithms MB09 and MBXI, in the particular environment of the blockchain.

The evolution of MB09 and MBXI – an introduction to MBXX

In 2020, I developed and patented another protocol that involves both MB09 and MBXI algorithms.

In my mind, one of the problems not wholly solved in Satoshi Nakamoto's paper was the *consensus problem*. Another issue (also noticed in MB09) is that we are dealing with a centralized system.

I wanted to overcome these problems, so I needed to implement a scheme such that the following conditions are met:

1. The protocol runs in a decentralized model.
2. The *consensus* for the validity of transactions is given by a mathematical deterministic function and not by a statistical probability of attack.

In other words, the problem of the *double-spending* of digital money has to be solved in a cryptographic way, and rather not with a *consensus* based on "game theory." Indeed, the consensus problem that Satoshi Nakamoto choose is a method based on the theory of the Byzantine Generals Problem.

This problem, explained through an informal description, involves a group of generals that obey a superior's order to attack an enemy. If the number of (generals = nodes) is more than three and some of them are dishonest, rebelling against the king's decision to attack, the attack could fail. In the following diagram, you can see the difference between a coordinated and an uncoordinated attack:

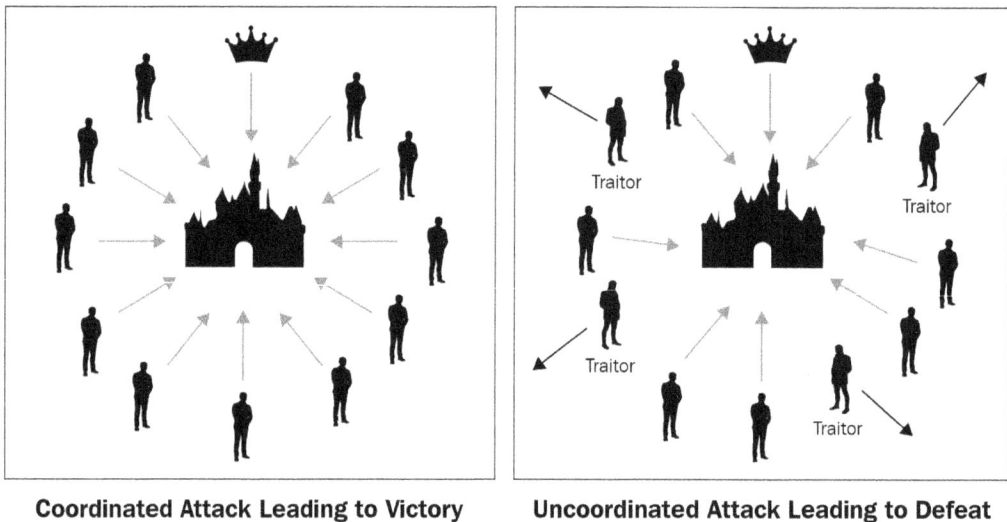

Coordinated Attack Leading to Victory   **Uncoordinated Attack Leading to Defeat**

Figure 6.11 – The Byzantine Generals Problem

Similarly, Satoshi Nakamoto relied on this theory when he supposed that the nodes of a system (the Bitcoin network), deputed to control the double-spending and the truthfulness of transactions, based on the *consensus problem*, could verify the truthfulness of the transactions. Satoshi Nakamoto supposed that the system could support an attack in which the *generals* would, *for the most part*, be honest.

Not everyone knows that in Bitcoin (as in most other cryptocurrencies, too), the deputed miners (through the *Proof of Work* to the validation of transactions) can't support an attack if the number of fidelity nodes is less than two-thirds of the total number of nodes involved. So, it's not enough that most of the nodes are trustable (51%), but the system requires, at the very least, 75% of the nodes to be trustable to be reliable.

It's rather strange that Satoshi Nakamoto, after having projected a very sophisticated protocol, mostly based on cryptography, to generate, spend, and give value to Bitcoin, switched to such a non-deterministic method to determine the *consensus* for the transactions.

Indeed, the *consensus* problem of Satoshi Nakamoto doesn't give mathematical evidence of a deterministic result. As I have said, it could only be valid if about 75% of the nodes inside of the system are trustable. Furthermore, if an attack occurs, the system will crash, and all the transactions will be invalid.

> **Important Note**
> *Proof of Work* is a method invented by Satoshi Nakamoto to produce and transact Bitcoins. It consists of solving a challenge based on SHA-256 (we have already explored SHA family Algorithms in *Chapter 4, Introducing Hash Functions and Digital Signatures.*

To avoid the problem of double-spending and maintain anonymous, peer-to-peer transactions, we have to rely on mathematical proof that gives a deterministic validity. I imagined a system in which a private blockchain can be governed and controlled only by computers. For example, an *autonomous decentralized organization* is almost completely managed by computers.

This concept of a **Decentralized Autonomous Organization (DAO)** was unknown until 2016. I use the concept of DAO as an organization that is self-regulated by a computer network or grid computing with high computation capacity, which is able to perform transactions and run programs to pursue a goal or multiple goals guided only by algorithms. Eventually, this is supported by artificial intelligence to make some decisions. The purpose is to create a decentralized organization, administrated by computers and not by humans. Or better, humans' role in such an organization could be to simply encode the software through the so-called *smart contracts*.

It doesn't matter who owns or is in charge of the hardware's maintenance. That is because this organization could be defined as a meta-infrastructure with no headquarters, no board of directors, and no limited time to survive, and, perhaps in the future, no human interference either. That's because hardware machines such as virtual machines exist in the cloud and run their algorithms in a virtual place where time and space don't have any real connotations. If we leave out the problems linked to such an organization's liability and the philosophical concept of creating an entity that is able to live eternally and is unstoppable, what remains is pure mathematics and logic. Theoretically, this scheme will be unstoppable because if it's well-developed, this organization could be split among a virtual internet organization that is completely agnostic to human power and regulated by algorithms.

Now, I will introduce this protocol and will give an overview of it. In the following diagram, I have represented a decentralized scheme connected through some large central star nodes:

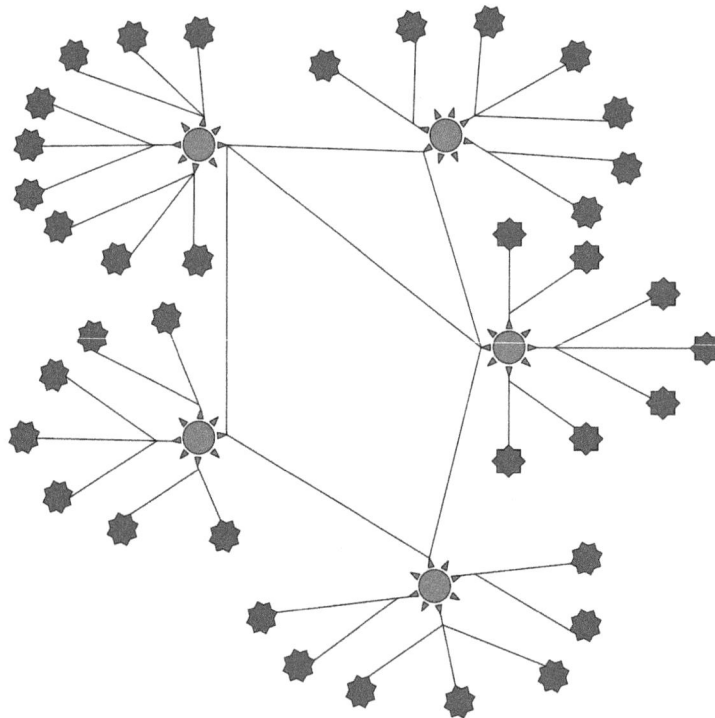

Figure 6.12 – A decentralized architecture

Now it's time to go through the steps of the MBXX protocol for digital money transactions in greater detail.

An explanation of the MBXX protocol

The first stage of this protocol is the initialization phase of all the parameters. It represents the *first-level* conditions.

The scheme consists of three steps, starting with an initialization stage of the algorithms adopted, parameters and keys.

Initialization stage:

All the parameters of the protocol are initialized. These parameters are as follows:

- The algorithms that are used in this protocol, such as (M/alg), (T/alg), and (ZK/Proof), for instance, SHA-256, MB09, RSA, D-H, MBXI, ZK13, zk-SNARK, and more.

- An algorithm that has been selected for the digital signature (this mostly depends on the encryption algorithm used).

- The parameters that are used in the algorithm, such as the private keys of the users, [a], [b], [c].... [n], and the correspondent public keys, (A, B, C, ..., N).

- The amounts of digital money detected by each account of the users from where the private keys have been derived.

- The generator, (g).

- The random numbers to generate ZK protocols, [k1, k2, ..., kn].

After this stage, we proceed with delineating the protocol.

Step 1: **Check the total balance with a meta-algorithm**:

A first-level algorithm or a *meta-algorithm*, represented by the (M/alg) notation, is deputed to check the initial balance and the balances generated after the transactions.

It is not necessary to have to use the system described in MB09 to perform the homomorphic balance; in fact, there could be something better than this. However, here, I will use the correspondence used in MB09 as a mere representation.

So, let's return to the extended Fermat's Last Theorem, as we know from MB09:

```
[a]^p + [b]^p + [c]^p+ … + [n]^p ≡ [z]^p (mod p)
A + B + C+ … +N = Z
```

This is (M/alg). It is made up of both the preceding equations and its role is double:

- It is used to check for the *isomorphic balance* (Z) in a certain instance of time.

- It is used to re-balance the system if new digital money is injected into the system (that is, the sum of the amount of digital money expressed in *isomorphic balance* during a certain time, t0, t1, ..., tn, changes the amount because of the initial balance (Z0) augmented if new digital money is injected).

In other words, the *isomorphic calculation* represents each term of the first preceding equation, [a^p, b^p, ..., n^p], transposed as being correspondent with (A, B, ..., N).

So, as we have already discovered, it will be very difficult for an attacker to recover [a] even if he knows its correspondent public value, (A). That is because this is a *one-directional* function. In this way, the balance of the accounts will be protected with an appropriate padding process. Additionally, you can decide to use another (M/alg) notation to obtain this correspondence, such as a *hash system*, where [a] corresponds to (A) ---> H[a], or another checksum system.

Here, you can discover that [a, b, c, ..., n] are the amounts of digital money at a determinate time: t0, t1, t2, ..., tn. Each of these elements corresponds to the [private amounts] of a person or of a computer, such as Alice, Bob, Carl, and Nate. Corresponding to any private parameters (the amounts of digital money), [a], [b], ..., [n], we have each public correspondent parameter: (A), (B), ..., (N). These public parameters represent the *isomorphic values* transposed by the hidden parameters in a public domain.

In other words, [a] corresponds to (A), [b] to (B), ..., [n] to (N).

Step 2: The transaction process:

Transactions among users take place at any point in time: t0, t1, ..., tn. You can figure out that here, [M] is a set digital amount of money, where [Ma, Mb, ..., Mn] are single amounts of digital money exchanged between the network participants at a particular time, t0, t1, ..., tn. For instance, we can think of [a0] = $10,000 as the initial balance of (A) and [Ma] = $1,500 as an amount of money transferred from the account of (A) to account (B).

The exchanging of [Ma], [Mb], ..., [Mn] between the users comes across as a cryptographic algorithm that I have called **Transmission/Algorithm (T/alg)**. By now, we assume that (T/alg) is well known and is made by a private/public key algorithm such as MBXI, for example.

So, after the first transaction performed, for instance, between (A) and (B) in times (t0) and (t1), the amounts of the single [a0] and [b0] accounts will be changed to [a1] and [b1]. For instance, if (A) transfers [Ma] = \$1,500 to (B) and the initial balance of A = \$10,000, then the single balance of (A) at time (t1) will be as follows:

```
(t0)  [a0]  =  10,000
(t1)  [a1]  =  a0-Ma
(t1)  [a1]  =  10,000 - 1,500 = $8,500
```

A digital signature has to be performed in (T/alg) in order to identify the sender to the receiver.

So, in this case, using MBXI, Alice adds her digital signature in this way:

$$Sa \equiv (Hma)^x \ (mod \ p)$$

So, (Sa) is the digital signature added by Alice, (Sa) is given applying the hash of the message (Hma) elevated to the encryption key [x], where (Hma) is given by the following function: (Hma) = Hash of [M].

And, as you remember from MBXI's signature, the private key [x] is given by solving the following equation:

$$\{ [K_A^b + e_B] \ (mod \ p) \} \ * \ x \equiv 1 \ \ mod \ (p-1)$$

Here, the (Sa) signature transmitted along with the cryptogram (c) will be verified by (B) in the opposite way:

$$(Sa)^\wedge y = (Hma) \ (mod \ p)$$

After Bob has verified that (Hma) corresponds to [M], he accepts the signature and receives the payment.

Step 3: The verification process:

This step makes all the transactions verified by the administrator of the system. All this process could be implemented by an autonomous organization similar to a DAO.

The system will operate the first validation on the *homomorphic balance* to check whether there is double-spending among the calculations. If this verification is okay, then after the transactions (made peer to peer through T/alg), the admin (or, better, the admins, because the system is computer decentralized) will proceed with a *blind check* of the accounts of the transmitter and the receiver to verify that the balance of the accounts will cover the amount of money transmitted. This verification will be done by applying a particular zero-knowledge protocol. I have called this algorithm **ZK/Proof**.

This ZK/Proof is able to detect whether a user tries to trick the system by sending a fake verification parameter in order to try to set up a double spending. In this case, for example, a user could try to make a payment but his balance doesn't cover the amount processed. This particular ZK/Proof recognizes the problem automatically and refuses the transaction, without any method of consensus based on a proof of work or other proofs to demonstrate. You can see a scheme of the first transaction and balance performed at time (t0/t1) with the MBXX protocol in the following diagram:

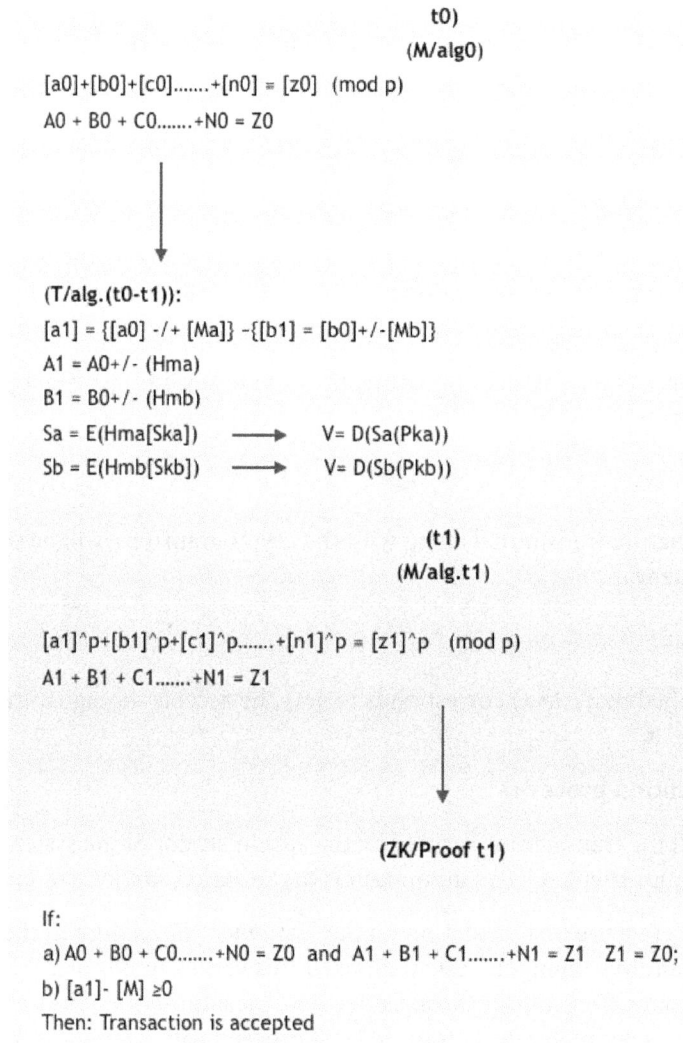

t0)

(M/alg0)

$[a0]+[b0]+[c0].......+[n0] = [z0]$ (mod p)

$A0 + B0 + C0.......+N0 = Z0$

$\downarrow$

(T/alg.(t0-t1)):

$[a1] = \{[a0] -/+ [Ma]\} -\{[b1] = [b0]+/-[Mb]\}$

$A1 = A0+/- (Hma)$

$B1 = B0+/- (Hmb)$

$Sa = E(Hma[Ska])$ $\longrightarrow$ $V= D(Sa(Pka))$

$Sb = E(Hmb[Skb])$ $\longrightarrow$ $V= D(Sb(Pkb))$

(t1)

(M/alg.t1)

$[a1]^\wedge p+[b1]^\wedge p+[c1]^\wedge p.......+[n1]^\wedge p = [z1]^\wedge p$ (mod p)

$A1 + B1 + C1.......+N1 = Z1$

$\downarrow$

(ZK/Proof t1)

If:

a) $A0 + B0 + C0.......+N0 = Z0$ and $A1 + B1 + C1.......+N1 = Z1$ $Z1 = Z0;$

b) $[a1]- [M] \geq 0$

Then: Transaction is accepted

Figure 6.13 – The scheme of the first stage of the MBXX protocol

If no other transactions are performed in (t1), then the protocol will go on with times (t2), (t3), and so on, given by a timestamp.

Notes on the MBXX protocol

If you're curious about this protocol, this section's scope offers more information, focusing attention on particular functions.

Referring to the preceding scheme, we find the equation: (T/alg(t0-t1)). This function expresses all the operations performed between the users at the time (t0 - t1).

> **Note**
>
> Pay attention to the - sign in the (t0 - t1) brackets. It is not a subtraction, but it means that we are moving from t0 to t1.

So, the function is as follows:

```
[a1] = {[a0] -/+ [Ma]} ~ {[ b1] = [b0]+/-[Mb]}
 A1 = A0 - (Hma) or A0 + (Hmb)
 B1 = B0 - (Hmb) or B0 + (Hma)
```

So, the secret amount, [a1], is generated by [a0] added or subtracted by amount [Ma], corresponding to amount [b1], generated by the transferred/received amount, [Mb]. Following this operation, the corresponding public parameter, (A1), is generated by the previous (A0) parameter subtracted by (Hma) if amount [M] is received by (B). The same goes for (B1), which is generated by the previous (B0) parameter added to or subtracted by the hash of the original amount transferred or received by (A).

This (Z0) scheme represents the amount of money in circulation in the system at a given instant, (t0). It doesn't change if no digital money is injected into the system. Instead, if a transaction takes place, it will change the corresponding users' values (parameters) involved in the transaction.

As we discussed in *step 2*, the *transaction process* uses any public/private key algorithm. It's easy to demonstrate that using MBXI, as the system works well. Because the amounts of digital cash are transferred directly from user to user, this can be considered as the effects of a peer-to-peer system, where no third party is involved in the transactions. So, theoretically, no financial or banking institution is required to perform the transactions.

A digital (S) signature is added by the sender and verified and accepted by the receiver if the decryption of the (V) signature returns the hash of [M] = (Hm). You can see in the scheme applied to the preceding diagram in the representation of (T/alg):

```
Sa = E(Hma[Ska]) --------> V= D(Sa(Pka))
Sb = E(Hmb[Skb]) --------> V= D(Sb(Pkb))
```

Here, (Sa) is Alice's signature, and (Sb) is Bob's signature.

The final step is the verification process, and it's related to a zero-knowledge protocol.

The first verification is as follows:

```
A0 + B0 + C0+ ... +N0 = Z0
A1 + B1 + C1+ ... +N1 = Z1
```

To be valid, the result must be Z1 = Z0.

This means that the public parameter, (Z1), which is the result of the *homomorphic balance* related to time (t1), is correspondent to (Z0) related to time (t0) even if the partial elements of the equation changed in value. This condition guarantees (if no money is injected into the system between times t0 and t1) that the transfer of money between the users into the networks is neutral or balanced.

In other words, the amounts of the transactions made by the users of the system are balanced. However, the verification of this condition by itself does not guarantee that no double-spending has been processed. In fact, it is possible that double-spending was performed if the parties operating in the system are untrustworthy.

Hence, here, we need to implement a *blind verification* or *homomorphic validation* to check that the single amounts of the accounts don't fall below zero. If this condition is also satisfied, then the system (DAO) will ensure that no double-spending has been performed.

Finally, we can say that the necessary condition to satisfy the system is a neutral homomorphic balance. The condition to be validated is that any single balance never falls below zero.

Conclusions on the MBXX protocol

This protocol should overcome the two problems identified at the beginning of this section:

1. The protocol runs in decentralized autonomous mode with no third parties involved in the transactions.

2. The *consensus* for the validity of transactions is given by a mathematical function and not by a statistical probability of attack.

In terms of the first problem, *decentralized autonomous mode*, I have already appointed the transactions to work through a cryptographic algorithm performed between the users in a peer-to-peer system. Moreover, the verifier (a computer) can only block a transaction if the verification process isn't complete. The verifier doesn't know how much money has been transferred or the number of single-money accounts. It only knows the original amount of money put into the system and whether a balance is misaligned.

In terms of the second problem, *consensus*, it can be demonstrated that a pure cryptographic model can perform it. This is based on two verifications: the *isomorphic balance* represented by (M/alg) and the **zero-knowledge proof** (**ZKProof**) that gives a blind comparison and states whether a single account balance is positive or negative, accepting only positive balances.

Summary

In this chapter, we analyzed some of my inventions. These algorithms and protocols were primarily implemented to cover some topics related to public/private encryption systems.

Among these protocols, we first saw MB09, which aimed to become a standard for secure digital payment transactions processed in the telecommunication field.

Then, we examined MBXI, a public/private encryption algorithm that was implemented as an alternative to RSA. We explored how it is possible to use a digital signature in this algorithm and the different signature methods: direct and with an appendix.

Moving on, we explored the last protocol: MBXX. This is an evolution of the MB09 and MBXI protocols to overcome the problem of double-spending and the so-called consensus problem. This protocol could be used in the future for a decentralized payment system as an alternative method to Proof of Work (or Proof of Stake or other additional proofs) proposed for the validation of transactions in cryptocurrency, based on statistical accuracy.

So, now you have learned about new methods and systems in public/private encryption, such as MB09, MBXI, and MBXX, and you have also become familiar with some schemes in centralized and decentralized crypto-systems related to the new era of digital currency.

These topics are important because you now have a deep understanding of complex cryptographic schemes. In the following chapters, we will explore many correlations with this part, particularly in *Chapter 9, Crypto Search Engine*, where I will explain the basics of the CSE system.

Now that you have learned about these new algorithms, we can explore another fascinating topic: elliptic-curve cryptography.

7
Elliptic Curves

Elliptic curves are the new frontier for decentralized finance. Satoshi Nakamoto adopted a particular kind of elliptic curve to implement the transmission of digital currency in Bitcoin called **secp256K1**. Let's see how it works and what the main characteristics are of this very robust encryption.

In this chapter, we will learn the mathematical basics of elliptic curve cryptography. This topic involves geometry, modular mathematics, digital signatures, and logic.

Moreover, I will present the special kind of elliptic curve implemented for the digital signature of Bitcoin known as secp256K1.

Finally, we will discuss the possibility of an attack on elliptic curves.

So, in this chapter, we will cover the following topics:

- The genesis of cryptography on elliptic curves
- Mathematical and logical basics of elliptic curves
- The Diffie–Hellman key exchange based on elliptic curves
- An explanation of ECDSA on secp256K1 – the digital signature of Bitcoin
- Possible attacks on elliptic curves

Let's dive deep into this intriguing topic, based on geometry and applied in cryptography.

An overview of elliptic curves

Around 1985, Victor Miller and Neal Koblitz pioneered *elliptic curves* for cryptographic uses. Later on, Hendrik Lenstra showed us how to use them to factorize an integer number.

Elliptic curves are essentially a geometrical representation of particular mathematical equations on the Cartesian plane. We will start to analyze their geometrical models in the 2D plane, conscious that their extended and deeper representation is in 3D or 4D, involving irrational and imaginary numbers. Don't worry now about these issues; it will become clearer later on in this chapter.

Elliptic Curves Cryptography (**ECC**) is used to implement some algorithms we have seen in previous chapters, such as **RSA**, **Diffie–Hellman** (**D–H**), and **ElGamal**.

Moreover, after the advent of the revolution in digital currency, a particular type of elliptic curve called secp256K1 and a digital signature algorithm called **Elliptic Curve Digital Signature Algorithm** (**ECDSA**) have been used to apply digital signatures to Bitcoin to ensure that transactions are executed successfully.

This chapter intends to take you step by step through discovering the logic behind elliptic curves and transposing this logic into digital world applications.

It has been estimated that using 313-bit encryption on elliptic curves provides a similar level of security as 4,096-bit encryption in a traditional asymmetric system. Such low numbers of bits can be convenient in many implementations requiring high performance in timing and bandwidth, such as mobile applications.

So, let's start to explore the basis of elliptic curves, which involve geometry, mathematics, and many logical properties that we have seen in earlier chapters of this book.

Operations on elliptic curves

The first observation is that an elliptic curve is not an *ellipse*. The general mathematical form of an elliptic curve is as follows:

```
E: y^2 = x^3 + ax^2 + bx + c
```

> **Important Note**
> E: represents the form of the elliptic curve, and the parameters (a, b, and c) are coefficients of the curve.

Just to give evidence of what we are discussing, we'll try to plot the following curve:

```
E: y² = x³ + 73
```

As we can see in the following figure, I have plotted this elliptic curve with WolframAlpha represented in its geometric form:

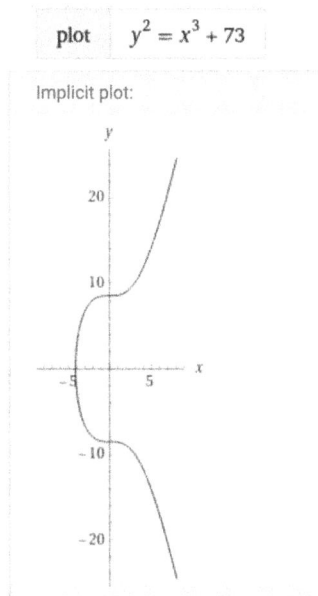

Figure 7.1 – Elliptic curve: E: y^2 = x^3 + 73

We can start to analyze geometrically and algebraically how these curves work and their prerogatives. Since they are not linear, they are easy to implement for cryptographic scopes, making them adaptable.

For example, let's take the curve plotted previously:

```
E : y^2 = x^3 + 73
```

When $(y = 0)$, we can see that, geometrically, the curve intersects the x axis at the point corresponding to $(x = -4.1793...)$; this is mathematically given if we substitute $y = 0$ into the equation:

```
y^2 = x^3 + 73
```

As a result, one of the three roots of the curve will be the cubic root of -73.

When (x = 0), the curve intersects the *y* axis at two points: (y = +/-8.544003...). Mathematically, we substitute (x = 0) into the equation:

```
y^2 = x^3 + 73
```

This obtains the intersection between the curve and the axis of *y*, as you can see in *Figure 7.1*.

Another curious thing about elliptic curves, in general, is that they have a point that goes to infinity if we intersect two points in the curve symmetrically with respect to the *y* axis. You can imagine the third point as an infinite point *lying* at the infinite end of the *y* axis. We represent it with O (*at the infinity point*). We will see this better later when we discuss adding points to the curve.

One of the most interesting properties of elliptic curves is the SUM value of two points of the curve.

As you can see in the following figure, if we have two points, P and Q, and we want to add P to Q, it turns out that if we draw a line between P and Q, a third point, -R, is given by the intersection between this line and the curve. Then, if you take a reflex of point -R with respect to the *x*-axis line, you find R; this is the sum of P and Q. It's easier to take a look at the following diagram to understand this geometrical representation of the SUM value:

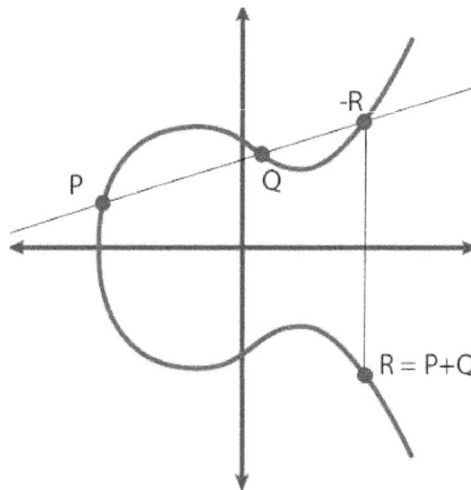

Figure 7.2 – Adding and doubling points on the elliptic curve

Now, let's look at how the addition point will be represented algebraically.

First, we take the coordinates of P and Q, and then calculate (s), as follows:

```
s = (yP - yQ)/(xP - xQ)
```

To compute xR, the *x* coordinate of point R, we have to perform this operation:

```
xR = s^2 - (xP + xQ)
```

To compute yR, the *y* coordinate of point R, we have to perform this operation:

```
yR = s(xP - xR) - yP
```

What about computing P + P = R = 2P, the so-called **point double**?

If we want to represent a P point added to itself so that it becomes 2P, the geometrical representation is given by the tangent passing through the P point and intersecting the curve at the R point, and again finding the reflexed point of the sum that is the symmetric point R, as shown here:

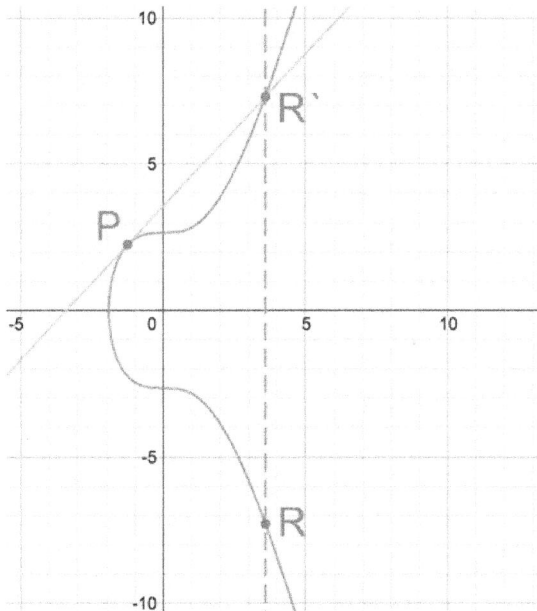

Figure 7.3 – 2P point double

Geographically, it is very similar to computing the sum of the point, as we saw in the preceding example.

We have to draw the tangent line in the P point, and we find the point of intersection between the line and the curve in the R' = (-2P) point; finally, the reflexed *x*-axis point on the curve will give the P double point, that is, R (2P).

What about computing P+P algebraically? In this case, (t) will be the following:

```
t = (3XP^2 + a)/2YP
```

Remember that (a) is a parameter of the curve.

So, to find the *x* coordinate of R, we have the following:

```
xR = t^2 - 2X_p
```

The equation for the *y* coordinate of R is as follows:

```
yR = t (xP - xR) - yP
```

There is one more case we need to address when we operate with elliptic curves: how to add vertical points.

Here is O represented by the *point of infinity*, given by the SUM value between P and Q if xP = xQ:

O (the point at infinity)

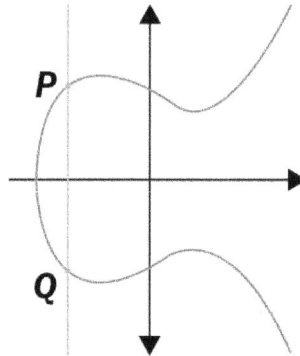

Figure 7.4 – Adding vertical points

Algebraically, the point at infinity is given as follows:

```
P + Q = O (point at infinity), if xP = xQ
```

Alternatively, it can be given like this:

```
P + P = O (point at infinity), if xP = 0
```

If we assume that A and B are the two points to add, we can summarize all the expressions into the following representation:

$$A = (x_A; y_A) \qquad B = (x_B; y_B)$$

$$x_C = \left(\frac{y_B - y_A}{x_B - x_A}\right)^2 - x_A - x_B$$

$$y_C = \left(\frac{y_B - y_A}{x_B - x_A}\right)(x_A - x_C) - y_A$$

2*A = A+A is obtained by the following equations:

$$x_{2A} = \left(\frac{3x^2_A}{2y_A}\right)^2 - 2x_A$$

$$y_{2A} = \left(\frac{3x^2_A}{2y_A}\right)(x_A - x_{2A}) - y_A$$

Figure 7.5 – Operations of addition and multiplication on elliptic curves

Now that we have seen adding and doubling points, let's see how to perform scalar multiplication, another important operation for cryptography on elliptic curves.

Scalar multiplication

At this point, we have to get familiar with another typical operation of elliptic curves, **scalar multiplication**. This process mathematically represents the sum between P and Q and makes it possible to calculate 2P, 3P, ..., nP on elliptic curves.

The logic behind scalar multiplication is not difficult, but it needs practice to become familiar with it.

Practically, let's solve a multiplication on the elliptic curve of the following form:

```
Q = n*P
```

This is called **repeated addition**, and it can be represented as follows:

```
Q = {P + P+ P + P...+ P} n-times
```

As we have to add the coordinates of a point to itself, we can figure out the scalar multiplication as the sum of a point with itself n times, so we have, for example, the following:

```
P + P = 2P
```

This, as we have seen, is the formula of double point (geometrically represented in *Figure 7.3*) transposed into an algebraic representation of SUM.

Always remember that we are dealing with a curve, so the coordinates of the new intersection point are as follows:

```
2P (X₂ₚ, Y₂ₚ)
```

For the same logic, we can go on with 3P, 4P, ... 7P:

```
R = 3P
R = 2P +P
```

Then we follow with 4P:

```
R = 4P
R = 2P + 2P
```

Then we follow with 5P:

```
R = 5P
R = 2p + 2P + P
```

Then we follow with 6P:

```
R = 6P
R = 2P + 2P +2P
```

Alternatively, it can be given like this:

```
R = 2(3P)
R = 2 (2P+ P)
```

Finally, we have the following:

```
R = 7P
R = P + (P + (P + (P + (P + (P + (P))))))
```

Alternatively, it can be given like this:

```
R = 7P
R = P + 6P
R = P + 2(3P)
R = P + 2(P + 2P)
```

It is called multiplication, but in reality, scalar multiplication is more of a *breakdown* of the R number in a scalar mode because, as you can see, step by step, it is going to be reduced to the minimal entity of P: so, for instance, 7P becomes a product of P and 2P.

Following this logic, if we have to multiply a K number with P, we have to break it down to obtain a sum of minimal elements (P+P) or a 2P double point that will compose the multiplication required. Because of the formulas, we rely on addition and multiplication, as we saw in *Figure 7.5*.

If we have 9P, for example, we will have the following:

```
9P = 2(3P) + 3P
9P = 2(2P + P) + 2P + P
9P = P + 2P + 2(2P+ P)
```

Now that we have understood this operation's logic, let's discover why these operations are so important to generate a cryptographic system on elliptic curves.

As we saw in the previous chapters (for example, D–H in *Chapter 3, Asymmetric Encryption*) about cryptography, we are looking for one-way functions. These particular functions allow it to be easy to compute in one direction but are very difficult to perform in the opposite sense. In other words, it isn't easy to reverse-engineer the result.

Similarly, in elliptic curves, we have to find a function such as a discrete logarithm that allows us to perform a *one-way* function.

We'll now define the discrete logarithm problem transposed on an elliptic curve.

It's stated that *scalar multiplication* is a one-way function.

Given an elliptic curve: E.

Known: P, Q is a multiple of P.

Find k: Such that Q = k * P.

It is a *hard problem to solve.*

This is called a discrete logarithm problem of Q to the P base, and it is considered hard to solve because to find k means to calculate very complex operations. I will proceed with an example to better understand what we are dealing with.

Example:

In the elliptic curve group defined by the following, what is the discrete logarithm k of Q = (4,5) to the P = (16,5) base?

$$y^2 = x^3 + 9x + 17 \text{ over the field } F_{23}.$$

One (naïve) way to find k is to compute multiples of P until Q. The first few multiples of P are as follows:

```
P = (16,5) 2P = (20,20) 3P = (14,14) 4P = (19,20) 5P = (13,10)
6P = (7,3) 7P = (8,7) 8P = (12,17) 9P = (4,5)
```

Since 9P = (4,5) = Q, the discrete logarithm of Q to the P base is k = 9.

In a real application, k would be large enough that it would be infeasible to determine k in this manner.

This is the fundamental principle behind the D-H algorithm implemented on elliptic curves that we will discover next.

Implementing the D-H algorithm on elliptic curves

In this section, we will implement the D–H algorithm on elliptic curves. We saw the D–H algorithm in *Chapter 3*, *Asymmetric Encryption*. You should remember that the problem underlying the D–H key exchange is the discrete logarithm. Here, we will demonstrate that the discrete logarithm problem could be transposed on elliptic curves too.

First of all, we are dealing with an elliptic curve (mod p). The **base point** or **generator point** is the first element in the D–H original algorithm represented by (g), and here we denote it by (G). Let's look at some elements to take into consideration:

- G: This is a point on the curve that generates a cyclic group.

  Cyclic group means that each point on the curve is generated by a repeated addition (we have seen point addition in the previous section).

- Another concept is the **order of G** denoted by (n):

  ```
  ord(G) = n
  ```

  The order of (G) = (n) is the size of the group.

  The order (n) is also the smallest positive integer [k], giving us the following:

  ```
  kG = O (Infinity Point)
  ```

- The next element to take into consideration is the h cofactor:

  h = (number of points on E)/n

  In other words, (h) can be defined as the number of points on E (mod p) divided by the order of the curve (n).

  h = 1 is optimal.

  If h > 4, the curve is more vulnerable to attacks.

Let's now analyze step by step how D–H transposed on E works.

Step 1: Parameter initialization:

Now let's look at the public shared parameters to initialize the D–H model on E (mod p):

```
{p, a, b, G, n, h}
```

p is the (mod p) as we have already seen in the original D–H algorithm.

a and b are the parameters of the curve.

G is the generator.

n is the order of G.

h is the cofactor.

Step 2: Crafting the shared key on E (mod p): [K]

After the parameter initialization, Alice and Bob will define the type of the curve to adopt among the family of E (mod p):

```
y^2 = x^3 + ax + b (mod p)
```

Bob picks up a [β] private key such that 1≤ [β] ≤ n-1.

Alice picks up a [α] private key such that 1≤ [α] ≤ n-1.

After choosing random keys, Bob and Alice can compute their public keys.

Bob computes the following:

```
B = [β] * G
```

Alice computes the following:

```
A = [α] * G
```

Now (like in the original D–H algorithm), there is a generation of shared keys exchanging the public parameters:

1. Bob sends B (xB and yB) to Alice.
2. Alice receives B.
3. Alice sends A (xA and yA) to Bob.
4. Bob receives A.
5. Bob computes the following:

    ```
    K = [β] *A
    ```

6. Alice computes the following:

    ```
    K = [α] * B
    ```

7. Finally, Bob and Alice hold the same information: the point on E: [K].

This point, [K], is the shared key between Alice and Bob.

Important Note

If Eve (the attacker) wants to know [K], she has to know [α] or [β], the private keys of Alice and Bob, given by the following functions:

B = [β] * G or A = [α] * G

To recover [K], Eve has to be able to solve the discrete logarithm on E (mod p), which, as we have seen, is a hard problem to solve.

A numerical example is as follows:

Let's assume that the domain of the curve is the following:

```
E: y^2 = x^3 + 2x +2 (mod 17)
```

You can take a look at this E plotted in *Figure 7.6*.

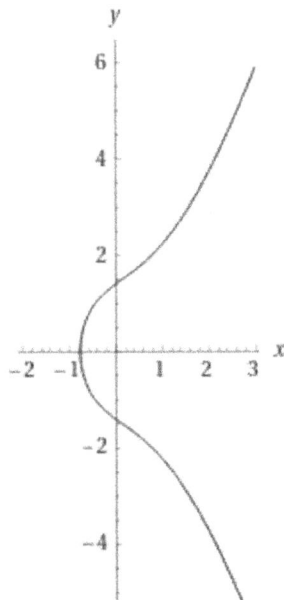

Figure 7.6 – The plotted curve used as an example to implement the D–H algorithm

The generator point is the following:

```
G (5,1)
```

First of all, we have to find the order of the n curve.

To do that, we should calculate all the points until we reach the minimum integer that allows the cycling group.

So, we start to calculate 2G.

Using the formula of the double point, we have the following:

```
t = (3X_p^2 + a)/2Y_p
t = (3*5^2 + 2)/2*1 = 77 * 2^-1 = Reduce[2*x == 77, x, Modulus
-> 17] = 13 (mod 17)
t = 13
```

Now that we have calculated (t), it's possible to use it to find the x and y coordinates on the 2G curve:

```
x2G = s^2 -2xG = 13^ 2- 2*5= Mod[13^2 - 2*5, 17] = 6 (mod 17)
y2G = s(xG - x2G) -yG = 13 (5-6) - 1 = -13-1 = -14 (mod 17) = 3
(mod 17)
```

So, we found the coordinates of 2G:

```
2G = (6,3)
```

Now, we should go on to compute 3G, 4G, ..., nG until we find the point of infinity, OG.

It is a lot of work to do it by hand, but it is also a good exercise to practice by yourself.

I will let you calculate all the scalar multiplications until the OG point, giving the results for some points:

```
G  = (5,1)
2G = (6,3)
3G = (10,6)
......
9G  = (7,6)
10G = (7,11)
```

This continues until it turns out that the point of infinity, OG, is the following:

```
19G = OG
```

This means that the order of G is the following:

```
n = 19
```

So, the parameters of E are the following:

```
G = (5,1); n =19
```

Bob picks up Beta:

```
Beta = 9
```

Alice picks up Alpha:

```
Alpha = 3
```

Bob calculates his public key (B):

```
B = 9G = (7,6)
```

Alice calculates her public key (A):

```
A = 3G = (10,6)
```

Alice sends (A) to Bob:

```
[β]*A = 9A = 9(3G) = 8G = (13,7)
```

Bob sends (B) to Alice:

```
[α]*B = 3B =3*(9G)= 8G =(13,7)
```

So, as you can see, the shared key is [K] = (13,7).

The parameters used in the example are too small to be implemented in a real environment. In reality, the numbers have to be larger than the ones I have used in the preceding example. However, this algorithm is computationally easier than the original D–H one and can be used with smaller parameters and keys.

However, we have to rely on curves that are well structured and architected by professional cryptographers and mathematicians.

> **Important Note**
>
> Implementing D–H on E (mod p) doesn't necessarily prevent MiM attacks, just like in the original D–H algorithm (as seen in *Chapter 3, Asymmetric Encryption*).

Now that we have more confidence with the elliptic curve and its operations, we can go through an interesting elliptic curve case, analyzing the algorithm adopted for the Bitcoin digital signature ECDSA.

Elliptic curve secp256k1 – the Bitcoin digital signature

ECDSA is the digital signature scheme used in Bitcoin architecture that adopts an elliptic curve called secp256k1, standardized by the **Standards for Efficient Cryptography Group (SECG)**.

ECDSA suggests (a = 0) and (b = 7) as parameters in the following equation:

```
E: y² = x³ + 7
```

For a more formal presentation, you can read the document reported by the SECG at `https://www.secg.org/sec2-v2.pdf`, where you can find the recommended parameters for the 256 bits associated with a Koblitz curve and the other bit-length sister curves.

This is the representation of secp256k1 in the real plane:

plot $y^2 = x^3 + 7$

Implicit plot:

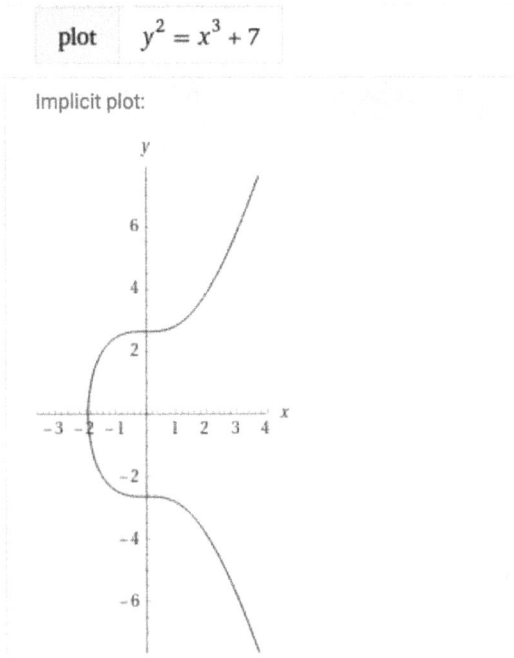

Figure 7.7 – secp256k1 elliptic curve

As we know, the elliptic curve has a part visible in the real plane and another representation in the imaginary plane. The form of an elliptic curve can be represented in 3D by a torus when the points are defined in a finite field, just as you can see in the following figure:

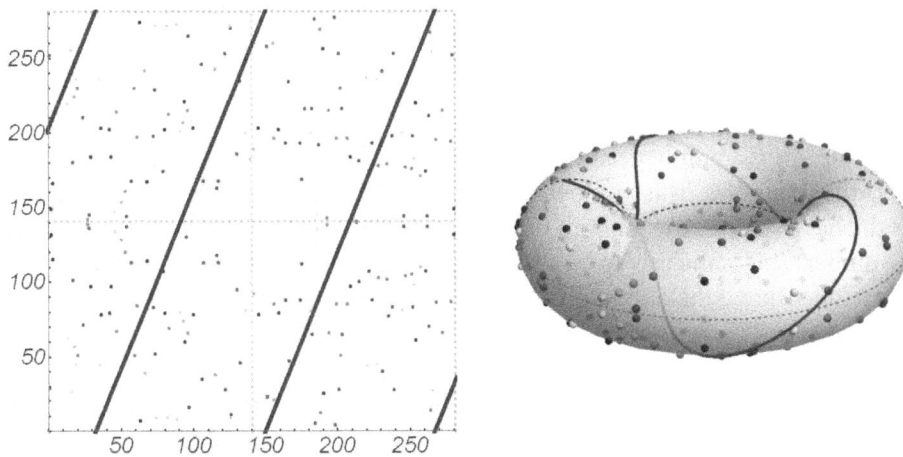

Figure 7.8 – 3D representation of an elliptic curve in a finite field

The secp256k1 curve is defined in the Z field as follows:

```
Z modulo 2^256 - 2^32 - 977
```

Or, written in a different way, this is how it looks as an integer:

```
115792089237316195423570985008687907853269984665640564039457584
007908834671663
```

In this, the coordinates of the points are 256-bit integers in a big modulo p.

The secp256k1 curve was rarely used before the advent of Bitcoin. As you can imagine, after it was used for Bitcoin, it became popular. Unlike most of her sisters, which commonly use a *random* structure, secp256k1 has been crafted to be more efficient. Indeed, if the implementation is optimized, this curve is 30% faster than others. Moreover, the constants (a and b) have been selected by the creator with a lower possibility of injecting a backdoor into it, which is different from the **National Institute of Standards and Technology** (**NIST**)'s other curves.

Please note that this last sentence is valid until proven otherwise.

Finally, in secp256k1, the generator (G), the order (n), and the prime (modulo p) are not randomly chosen but are functions of other parameters. All that makes this curve one of the best you can implement and it's useful for the scope of digital signatures. This is the reason why Bitcoin's developers chose it. In Bitcoin, precisely, we will see how many digits will be selected for the modulo (p), the order (n), and the base point (G) to make secp256k1 secure.

Let's go on now to explore how the digital signature ECDSA algorithm is implemented in secp256k1.

Step 1 – Generating keys

The [d] private key is a number randomly chosen within the range $1 \leq k \leq n-1$, where (n) is the order of G. This number has to be of a length of 256 bits so that it computes the hash value of the random number with SHA-256, which gives as output a 256-bit number. This is the way to be sure that the number is effectively 256 bits in length. Remember that if the output gives a number lower than $(n-1)$, then the key will be accepted; if not, it will make another attempt.

The public key (KPub) is derived by the following equation:

```
KPub = Kpriv*G
```

Thus, the number of possible private keys is the same as the order of G: n.

To calculate the public key (Q), starting from the private key, [d], in secp256k1, we have to use the following equation:

```
Q ≡ [d] * G (mod p)
```

The result of the equation will be Q (the public key).

After this operation, we can switch to calculating the digital signature. Suppose that Alice is the sender of the signature and her [d], (G), and (Q) parameters have already been calculated. In a real case, Victor is the verifier (the miner) who has to verify the true ownership. Let's go on to see how to sign a Bitcoin transaction in secp256k1.

Step 2 – Performing the digital signature in secp256k1

Remember that to sign a document, [M], or, even better in this case, a value in Bitcoin, [B] (as we discussed in *Chapter 4*, *Introducing Hash Functions and Digital Signatures*), it is always recommended to compute Hash[B] = z.

So, we will sign (z), which is the hash of the message, and *not* [M] directly.

Another parameter that we need is [k], which is an ephemeral key (or session key).

The sub-steps to perform the digital signature (S) are similar to those for a public/private key algorithm. The difference here is that the discrete logarithm is obtained through a scalar multiplication:

1. Alice chooses a random secret, [k], which gives us the following: [1≤ k ≤ n-1].
2. She calculates the coordinates, R(x,y) = k*G.
3. Alice finds r ≡ x (mod n) (the *x* coordinate of R(x,y)).
4. Alice calculates S ≡ (z + r*d)/ k (mod p) (this is the digital signature).

Alice sends the (r and S) pair to Victor for verification of the digital signature.

Step 3 – Verifying the digital signature

Victor receives the (r and S) pair. He can now verify whether (S) has really been performed by Alice.

To verify (S), Victor will follow this protocol:

1. Check whether (r) and (S) are both included between 1 and (n-1).

2. Calculate w $\equiv$ S^(-1) (mod n).

3. Calculate U $\equiv$ z * w (mod n).

4. Calculate V $\equiv$ r * w (mod n).

5. Compute the point in secp256k1: R (x,y) = UG + VQ.

6. Verify that r $\equiv$ Rx (mod n).

If (r) corresponds to Rx (mod n) – the x coordinate of the R point – then the verifier accepts the signature (S) corresponding to the value of Bitcoin, [B], the owner of which claims to own the amount of Bitcoin declared.

Without the support of a practical example, it is rather difficult to understand such a complex protocol. Some of the operations performed on secp256k1 for the digital signature of a Bitcoin transaction are quite complex to interpret and need a practical example to understand.

A numerical exercise on a digital signature on secp256k1

In this section, we will deep dive into the digital signature of secp256k1 in order to understand the mechanism behind the operations of implementation and validation of the digital signature.

Suppose for instance that the parameters of the curve are the following:

```
p = 67 (modulo p)
G = (2,22)
Order n = 79
Private Key: [d]= 2
```

So, as we have chosen a very simple private key, it is just enough to perform a double point to obtain the public key (Q):

```
Q = d*G
```

In this case, we proceed to calculate the public key (Q). First, we will be using the following formula to calculate the double point:

```
t = (3XP^2 + a)/ 2YP
t = (3*2^2 + 0)/ 2*22= 12/ 44 = Reduce[44*x == 12, x, Modulus
-> (67)] = 49
t = 49
```

Calculating Q = d * G using numbers implies replacing [d] and (G) with numbers:

```
Q = 2 * (2,22)
```

Relying on the formula of the double point, let's find the coordinates of Q (x and y), starting with x:

```
xQ = t^2 - 2XG
xQ ≡ 49^2 - 2 * 2 = 52 (mod 67)
xQ = 52
```

In the same way, let's find the y coordinate of yQ:

```
yQ = t(xG - xQ) - yGx
yQ ≡ 49 (2 -52) - 22 = 7 (mod 67)
yQ = 7
```

So, the coordinates of the public key (Q) are as follows:

```
Q (x,y) = (52, 7)
```

Now, Alice can perform her digital signature (S).

First, Alice computes the H [M] = z hash.

Suppose, z = 17 is the hash of the [B] value of the Bitcoin.

Alice chooses a random secret number for the [k] session key:

```
k = 3
```

Alice calculates the R (x,y) = k*G coordinates:

```
k*G = 3 * (G) = G + 2G
```

We have already calculated 2G = (52,7), so it's possible to rely on the additional formula to calculate G + 2G = (2,22) + (52,7).

Let's recall the formulas of point addition on elliptic curves:

```
t1 = (yQ-yG)/(xQ-xG)
```

To compute xR (the *x* coordinate of the R point), we have to solve this operation:

```
t1 ≡ (7 - 22)/(52 - 2) (mod 67)
t1 ≡ 60 (mod 67)
```

We are seeking [xR]:

```
xR ≡ t1^2 - (xG + xQ) (mod 67)
xR ≡ 60^2 - (2+52) = 62 (mod 67)
```

Let's consider the *x* coordinate of R:

```
r = 62 (mod 79) = 62
```

At this point, we can perform the signature applying the formula:

```
S ≡ (z + r*d)/k (mod p)
S ≡ (17 + 62 * 2)/3 (mod 67) = 47
```

We have gained a very important point – the signature (S):

```
S = 47
```

Alice sends the (S = 47, r = 62) pair to Victor.

To verify the digital signature, Victor receives (S, r) = (47, 62).

The public parameters that Victor has available are as follows:

```
z = 17
n = 79 (order of G)
G = (2,22) Base Point
Q = (52,7) Public Key
```

First of all, Victor has to check the following:

```
1≤ (47, 62) ≤ (79-1)
```

In fact, here we go – the signature passes the first check.

Victor now calculates (w), the inverse of the digital signature (S):

```
w ≡ S^(-1) (mod n)
```

That, as we have already seen on several occasions, means to perform the inverse functions of S (mod n):

```
Reduce[(S)*x == 1, x, Modulus -> (n)]= Reduce[(47)*x == 1, x,
Modulus -> (79)]= 37
w = 37
```

Then Victor calculates (U):

```
U ≡ z * w (mod n)
U ≡ 17 * 37 (mod 79)
U = 76
```

Now Victor is able to verify the signature:

```
V ≡ r * w (mod n)
V ≡ 62 * 37 (mod 79)
V = 3
```

The game is not finished yet; we have to *convert* through scalar multiplication the coordinates of V = 3 on secp256k1.

To do that, we have to perform the following equation:

```
R(x,y) = U*G + V*Q
```

We will split the preceding equation into two parts: (UG) and (VQ). We start by calculating UG:

```
U*G = 76G
    = 2*38G
    = 2*(2*19G)
    = 2*(2(G +18G)
    = 2*(2 (G + 2(9G)))
    = 2(2(G + 2(G +8G)))
    = 2(2(G + 2(G + 2(4G))))
    = 2(2(G + 2(G + 2(2(2G)))))
```

In order to reduce 76G through scalar multiplication, we have to perform six-point double (G) operations and two-point additions.

This shortcut will come in handy when the numbers are very large. There is *no* efficient algorithm able to perform such an operation of reduction like this, so we have to use our brain.

We already know the results of 2G = 2 (2,22) = (52,7), so we can reformulate the preceding operations, 2*(52,7) = (21, 42), already calculated by me, and we arrive at this further reduction:

```
UG = 2 *((52,7) + 2((2,22) + (21,42)))
   = 2 *((52,7) + 2(13,44)))
   = 2(2(38,26))
   = 2(27, 40)
   = (62,4)
UG = (62,4)
```

In the next step, we have to calculate VQ = 3Q = Q+ 2Q.

One method to perform this scalar multiplication is first to calculate 2Q = 2 *(52,7) = (25,17).

Then we proceed to compute Q + 2Q = (52,7) + (25,17):

```
VQ = (52,7) + (25,17) = (11,20)
VQ = (11,20)
```

Finally, we can add UG + VQ in a unique adding point:

```
R (x,y) = U*G + V*Q
R (x,y) = (62,4) + (11,20)
R (x,y) = (62,63)
```

For the last verification, Victor can check the following:

```
r ≡ Rx (mod n)
```

In fact, it turns out as follows:

```
r = 62 = Rx = 62 (mod 79)
```

So, finally Victor accepts the signature!

We have seen how complex it is even if we use small numbers to delve inside this protocol, as we have demonstrated in this example. So, you can imagine the enormous complexity that must be involved if the parameters are 256 bits or more. This elliptic curve is proposed to protect the ownership of Bitcoin. The signer, through their signature (S), can demonstrate they own the [B] value corresponding to the computed hash (z).

It's verified that secp256k1 is a particular elliptic curve, as we saw at the beginning of this section when I mentioned the characteristics of this curve, which, being different from others, is more efficient and has parameters chosen in a particular way to be implemented.

If you are curious about the implementation and the digital signature ECDSA algorithm, you can visit the NIST web page and the **Institute of Electrical and Electronics Engineers (IEEE)**.

NIST has published a list of different kinds of ECC and the recommended parameters of the relative implementations. You can find the document at this link: `https://nvlpubs.nist.gov/nistpubs/SpecialPublications/NIST.SP.800-78-4.pdf`.

IEEE is a private organization dedicated to promoting publications, conferences, and standards. This organization released the IEEE P1363-2000 (Standards Specification for Public-Key Cryptography), where it is possible to find the specifications to implement the ECC.

In the next section, you will find an attack against EDCSA private keys made by a hacker group calling itself *fail0verflow*, which announced it had recovered the secret keys used by Sony to sign in to the PlayStation 3. However, this attack worked because Sony didn't properly implement ECDSA, using a static private key instead of a random one.

Attacks on EDCSA and the security of elliptic curves

This attack on ECDSA can recover the private key, [d], if the random key (ephemeral key), [k], is not completely random or it is used multiple times for signing the hash of the message (z).

This attack, implemented to extract the signing key used for the PlayStation 3 gaming console in 2010, recovered the keys of more than 77 million accounts.

To better understand this disruptive attack (because it will recover not only the message but also the private key, [d]), we will divide it into two steps. In this example, we consider the case when two messages, [M] and [M1], are digitally signed using the same private keys, [k] and [d].

Step 1 – Discovering the random key, [k]

The signature (S = 47) generated at the time (t0) from the hash of the message, [M], as we know, is given by the following mathematical passages:

```
S ≡ (z + r*d )/k (mod p)
```

Here it is presented in numbers:

```
S ≡ (17 + 62 * 2)/3 (mod 67) = 47
S = 47
```

Suppose now we know z1 = 23 (the hash of the second message, [M1]) transmitted at the time (t1) and generating the signature (S1). Moreover, we are supposed to know the second signature (S1) given by the equation:

```
S1 ≡ (z1 + r*d)/k (mod p)
```

Using the Reduce function of Mathematica, we have the following result for (S1):

```
[k*x = (z1 + r*d), x, Modulus -> (n)] = 49
S1 = 49
```

Given (S - S1)/(z-z1) ≡ k (mod n).

Substituting the parameters in the preceding equation with the numbers of our example, we can easily gain [k] using the Reduce modular function of Mathematica, as follows:

```
Reduce[(47 - 49)*x == (17 - 23), x, Modulus -> (79)]
k=3
```

After we get [k], we can also gain the private key, [d].

Let's see what happens in the next step.

Step 2 – Recovering the private key, [d]

After we have recovered the random key, [k], we can easily compute the private key, [d].

We know that (S) is given by the following equation:

```
S = (z + r*d )/k
```

Switching (k) from denominator to enumerator at the left side, we can write the equation in the following form:

```
S*k = z + r*d
```

From knowing (k) = 3, we can find [d] because also all the other parameters (S,z,r) are public.

So, we can find [d] by writing the previous equation as follows:

```
d = (S*k -z)/r
```

All the parameters on the right side are known.

We can write it out in numbers, as follows:

```
(47 * 3 - 17)/r = 2
```

That is the number of the private key, [d] = 2, discovered!

Analyzing this attack, we come to a question: what happens if someone can find a method to generate multiple signatures (S1, S2, and Sn) starting from (z1, z2, and zn) or generate multiple signatures starting from (z)?

In other words, as (z) comes from the hash of the message (the value of Bitcoin recovered in a wallet) that we have called [B], what happens if we can generate an S1 signature starting from (z)?

I invite you to think about this question because if that is possible, then it should be possible to generate multiple signatures on a single transaction, all verified by the receivers.

Let's now see a brief analysis of ECC's computational power compared to other encryptions. Elliptic curve efficiency compared with the classic public/private key cryptography algorithm is shown in the following table. You can immediately perceive the lowest key size used in elliptic curves compared to the RSA/DSA algorithms. The equivalence you see in the following figure also takes into consideration the symmetric scheme key size (**Advanced Encryption Standard** (**AES**), for example). This scheme is undoubtedly shorter than the elliptic curve, but here is a comparison because, as you already know, the elliptic curve is able to spread out digital signatures differently by symmetric scheme encryption, so the effective comparison has to be done with asymmetric encryption:

Comparable Key Sizes for Equivalent Security

Symmetric scheme (key size in bits)	ECC-based scheme (size of n in bits)	RSA/DSA (modulus size in bits)
56	112	512
80	160	1024
112	224	2048
128	256	3072
192	384	7680
256	512	15360

Figure 7.9 – William Stallings' table of comparison – ECC versus classical cryptography

As you can see, 256-bit encryption performed on ECC is equivalent to a 3,072-bit key on RSA. Moreover, we can compare a 512-bit key on elliptic curve to a key size on RSA of 15,360 bits. Compared with symmetric encryption (AES, for example), ECC needs double the amount of bits.

> **Important Note**
> A warning regarding the key's length: it doesn't matter how long the modulus is or how big the key size is – if an algorithm logically breaks out, nothing will repair its defeat.

Considerations about the future of ECC

Now that we have seen how a practical attack on ECDSA works, one of the most interesting questions we should ask for the future is the following:

Is elliptic curve cryptography resistant to classical and quantum attacks?

At a glance, the answer could be that most elliptic curves are not vulnerable (if well implemented) to most traditional attacks, except for the same ones we find against the classic discrete logarithm (such as Pollard Rho or a birthday attack) and man-in-the-middle attacks in D–H ECC.

In the quantum case, however, Shor's algorithm can probably solve the elliptic curve problem, as we will see in the next chapter, dedicated to quantum cryptography.

Thus, if someone asks: are my Bitcoins secured for the next 10 or 20 years? We can answer: under determinate conditions, yes, but if the beginning of the quantum-computing era generates enough qubits to break the classical discrete logarithm problem, it will probably break the ECC discrete logarithm problem in polynomial time too. So, I agree with Jeremy Wohlwend (a Ph.D. candidate at MIT), who wrote about *Elliptic Curve Cryptography: Pre and Post Quantum*:

> *"Sadly, the day that quantum computers can work with a practical number of qubits will mark the end of ECC as we know it."*

Summary

In this chapter, we have analyzed some of the most used elliptic curves. We have seen what an elliptic curve is and how it is designed to be used in cryptography.

ECC has algorithms and protocols designed mainly to cover secrets related to public/private encryption systems, such as the D–H key exchange and the digital signature.

In particular, we analyzed the discrete logarithm problem transposed into ECC, so we have familiarized ourselves with the operations at the core of ECC, such as adding points on the curve and scalar multiplications.

These kinds of operations are quite different from the addition and multiplication we are familiar with; here, indeed, lies the strength of elliptic curves.

After the experimentation done on D–H ECC, we analyzed in detail secp256k1, which is the elliptic curve used to implement digital signatures on Bitcoin protocol through the ECDSA.

So, now that you have learned about elliptic curves and systems as alternative methods in public/private encryption, you can understand that one of the best properties of these curves is the grade of efficiency in their implementation, which can be a lower key size.

At the end of this chapter, we asked a question about ECC's robustness regarding quantum attacks. It was with this answer that I introduced the topic of *Chapter 8, Quantum Cryptography*. This chapter will, for sure, cover one of the most intriguing and bizarre topics of this book. Quantum computing and quantum cryptography will be the new challenge for cryptography's future.

8
Quantum Cryptography

In this chapter, I will explain the basics of **quantum mechanics** (**Q-Mechanics**) and **quantum cryptography** (**Q-Cryptography**). These topics require some knowledge of physics, modular mathematics, cryptography, and logic.

Q-Mechanics deals with phenomena that are outside the range of an ordinary human's experience. So, it might be difficult for most people to understand and believe this theory. First, we will start by introducing the bizarre world of Q-Mechanics and then examine Q-Cryptography, which is a direct consequence of this theory. We will also analyze quantum computing and Shor's algorithm before looking at which algorithms are candidates for post-Q-Cryptography.

So, in this chapter, we will cover the following topics:

- Introduction to Q-Mechanics and Q-Cryptography
- An imaginary experiment to understand the elements of Q-Mechanics
- An experiment on quantum money and quantum key distribution (BB84)

- Quantum computing and Shor's algorithm

- The future of cryptography after the advent of quantum computers

Let's dive deep inside this fantastic world, based on strange and (sometimes) incomprehensible laws.

Introduction to Q-Mechanics and Q-Cryptography

So far in this book, the kind of cryptography we have analyzed always followed the laws of logic and rigorous mathematical canons. Now, we are approaching it with a different logic. We are leaving the logic of classical mathematics and landing in a new dimension where something could be everything and the opposite of everything.

We have to face off with a kind of science that Niels Bohr (one of the fathers of Q-Mechanics) said this about: *"Anyone who can contemplate quantum mechanics without getting dizzy hasn't understood it."* Einstein's divergence defined the entanglement theory as a *phantasmatic theory*.

A couple of preliminary considerations about Q-Cryptography and Q-Mechanics are as follows:

- All the cryptography you have learned until now, even the most evolute, robust, and sophisticated, relies on two elements:

  I.   The difficulty to break it depends on the algorithm and its underlying mathematical problem: factorization, discrete logarithm, polynomial multiplication, and so on.

  II. The size of the adopted key. In other words, the length of the key or the field of numbers that the algorithm operates in to generate the keys defines the grade of computational break-even under which any algorithm fails. For example, if we define a public key (N) in RSA as N = 21, even a child in elementary school will be able to find the two secret numbers (private keys), p = 3 and q = 7, given by the factorization of the number; that is, 21 = 3*7. Conversely, if we define a field of operations of 10^1000, then the factorization problem becomes a hard problem to solve.

- In Q-Cryptography, the problem doesn't rely on any of these elements. There isn't any *mathematically hard problem to solve* that underlies this kind of cryptography (in the classical sense) but everything is based on a particular, even bizarre, hypothesis: in Q-Mechanics, an element (a particle, such as a photon) can be in a state of 1 and 0 at the same time. It can be present in two places at the same time. It is impossible to determine the state of a particle until it's measured. As we'll learn shortly, the concept of *time* itself in Q-Mechanics loses its meaning because the causality property is not an option, just a mere possibility.

Before we analyze the hypothesis that was anticipated in this preamble, let's start with an experiment that changed the way that people thought about microparticles.

The first experiment that changed the history of the science in this branch forever was created by Thomas Young, a polyhedric scientist, at the end of the 18th century. Besides that, Young was the first person to decode some hieroglyphics, so we can also count Young among cryptographers. He often used to walk beside a small lake where groups of ducks swam. He noticed how the waves that were created by the ducks while they swam interacted with themselves and intersected with each other, generating ripples. This reminded him of the same behaviors of the waves of light, which inspired his work.

The experiment, as shown in the following diagram, demonstrates that light waves behave like particles and cause multiple small stripes on the screen on the back of two slits where the light enters, instead of generating only two large strips.

So far, nothing paradoxical is hidden behind this experiment, except that if we run the experiment by shooting only one photon (a nanoparticle of light) every time (let's say, one each second) through the two slits, it turns out that the final effect at the back of the screen is not two dark lines (as we would imagine it to be) in correspondence with the two slits, but a long row of striped lines, just like the waves that are generated by light.

Here, you can see Young's experiment reproduced:

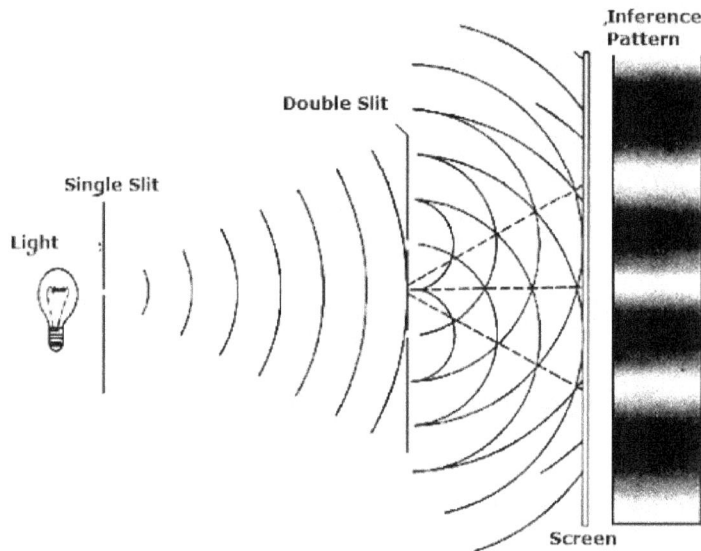

Figure 8.1 – Young's light experiment

In the next section, I will provide a variant of this experiment, reinvented for you to understand Q-Mechanics (or more likely to get you dizzy).

An imaginary experiment to understand the elements of Q-Mechanics

In this section, we will try to go deeper by providing an imaginary experiment to introduce Q-Mechanics.

This experiment's elements are similar to Young's experiment: a wall with two slits and a screen on the back. Later on, step by step, we will address more elements that will help you set up the experiment and understand the paradoxes of Q-Mechanics.

Suppose we shoot a photon toward the two slits shown in the following diagram. We will think of the photons as small marbles.

Step 1 – Superposition

Let's start by shooting marbles that are a normal size. Each time a marble is shot, it will likely move towards the slit and pass through it. So, let's assume that at the end of the first round, the result will be that most of the marbles will hit the screen in a straight line behind the slits:

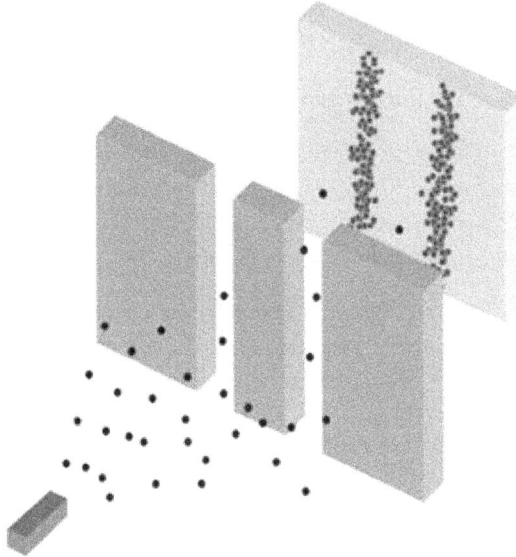

Figure 8.2 – The marbles hitting the screen, causing two separate lines at the back

Now, suppose that the two slits become very narrow and the marbles become very small. This is the case when we are dealing with photons, and we will see that the result is very different. Different from the result of normal marbles, a striped pattern is produced with marbles that are the same dimension as a photon. Take a look at the following diagram:

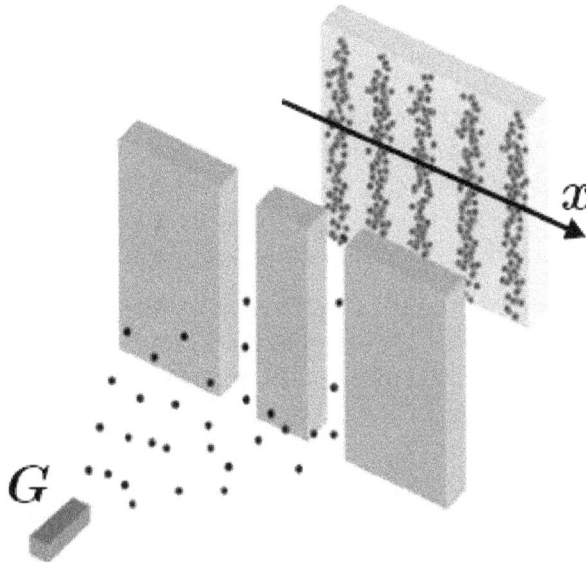

Figure 8.3 – The marbles are very small and cause a striped pattern

As we can see, everywhere in the universe, particles produce a striped pattern, as shown in the preceding diagram, provided that both the slits and the particles are small enough.

The result of this phenomenon is known as **waves**.

When a wave passes through a slit, it spreads out across the other side. If there are two slits instead of one, then two waves are produced that interact with themselves. These waves will produce the effect of a striped pattern, as we can see on the screen that was reproduced in Young's experiment, as shown in *Figure 8.1*.

Now, a question comes to mind: *if all the objects of the universe behave like waves, why do we only notice the stripes when we deal with small objects (marbles, like photons) and not when they are large?*

But the real problem with this representation is that if we only launch one marble (particle) each time, it still produces a striped pattern. The reason for this is related to the volume of energy related to the particles. So, how is it possible for a single marble (however small it may be) to pass simultaneously through the two slits?

Let's stop just for a second to reflect on this first important property. If we give the state of (0) to the photon that passes through the left slit and (1) to the photon that passes through the right slit, then we can say that the two photons are passing simultaneously in the two slits, so they are in a state of (0) and (1) simultaneously. This is the first important property in Q-Mechanics and it is called **superposition**.

If we block one slit and shoot the marbles just through one slit, the pattern that's generated will be the same as a normal marble hitting the screen, just behind the open slit directly. But if we unlock both slits, the striped pattern returns.

Superposition can offer a vision in which a particle can be in different states in a determinate moment. However, another part of physics, and some scientists, have a second interpretation of this experiment. This interpretation lies not in superposition but the equally *bizarre concept* of the *multiverse*. This concept means that many universes represent the different states that a particle is in.

This alternative vision of the multiverse also offers a philosophical theory of quantum reality that's transposed in our lives. For example, there could be different copies of myself who are playing different roles in different parallel worlds. Anyway, given the scope of this book, we will continue to experiment and talk about another important characteristic of Q-Mechanics: indetermination.

Step 2 – The indetermination process

Now, let's test what happens if we put a detector in front of each slit. We should expect both the detectors to reveal the marbles simultaneously because, as we saw earlier, shooting only one particle makes it appear in a superposition state; however, this is not what happens here.

If we shoot a marble small enough to be compared to a small particle, we will see that each marble only passes through one of the two detectors, never both. Any attempt to discover which of the two slits the marble passes through generates *indecision*, which causes the end of the striped pattern. So, this attempt forces the marble to pass through one slit, not both.

That is another fundamental property of Q-Mechanics: when the photons are in a state of superposition (or are playing in a multiverse), they are (0) and (1) at the same time; instead, when they are observed, the photons change their behavior and choose to be (0) or (1).

You can understand that the indetermination property is strictly correlated to superposition. We will return to this later, but for now, let's continue with our experiment.

There is a clarification we must make before continuing. If we close our eyes and don't observe the marbles passing through the slits, even if the detectors are put in front of the slits, the superposition state is still activated. Only when we open our eyes and look is the state of indetermination stopped and the marble is forced to take a position of (0) or (1). This means that the marble we are dealing with should be different from waves. Waves produce their effects in any condition, both if we observe or do not observe the experiment. This problem produces another interesting implication in physics, relating to philosophy, but that is beyond the scope of this book.

Now, suppose we put two objects behind the two slits. While shooting the marble through the slits, we can imagine that the marble hits one of the two objects. This is true when we look at the situation, but again, if we close our eyes and don't look, then the probability that the marble knocks down one or the other object is the same. In other words, the two objects can be compared to the famous thought experiment of **Schrödinger's cat**, where the cat is alive and dead at the same time until we discover it.

According to the mathematics that describes the probability of waves, neither outcome is certain. Only when we open our eyes and look can we know whether an object is alive, (1), or dead, (0). Moreover, the fact that we can only assign a certain *probability* to a particle situated in a certain place in the universe and that it comes out only when someone looks at it is logic that's very hard to understand and believe. However, despite its crazy logic, all the experiments we've looked at until now demonstrate that the indetermination process is valid! Probability is another fundamental element that we have to keep in mind when we deal with Q-Mechanics. The probability of objects being in one state or another and one place or another at the same time induced Einstein to be a strong adversary of the quantum theory and, in particular, of Schrödinger. Regarding this logic, Einstein commented, *"God does not play dice with the Universe."*

It's spin that we refer to when we talk about probability in Q-Mechanics. So, let's explore what spin is and why it is so important for our studies.

Step 3 – Spin and entanglement

To explain how a particle could be in two states simultaneously in Q-Mechanics, we need to introduce **spin**.

In literature, spin refers to the total angular momentum or intrinsic angular momentum of a body.

The direction of the spin of a particle can be described by an imaginary arrow crossing the particle in different directions. Particles spinning in opposite directions will have their arrows pointing in opposite directions, as shown in the following diagram:

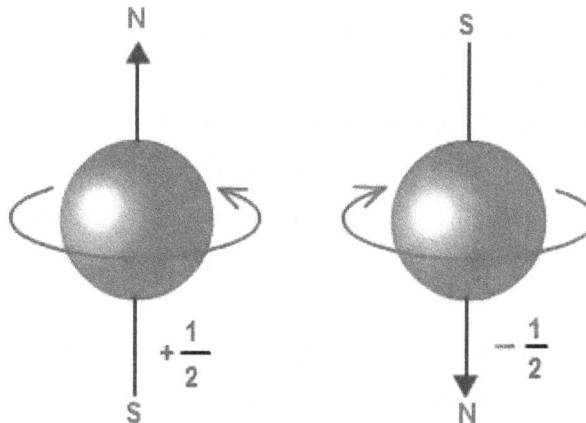

Figure 8.4 – Spin

Spin is very important for Q-Cryptography, as you will understand later when we analyze how to quantum exchange a key in more detail. The content or information that a particle can carry depends on its spin. This depends on the information that's enclosed in the spin direction; that is, whether Alice and Bob can determine the bits of a quantum distributed key. But we will look at this later.

Now, let's return to the concept of quantum information: we know that information in the normal world is made up of letters and numbers, but we can give or receive pieces of information in a discrete way, such as YES or NO, UP or DOWN, or TRUE or FALSE. We can also *discretize the information*, reducing it to a bit in the real world. We can do the same in Q-Mechanics, but we call it a **quantum bit** or **qubit**.

Let's look at an example of a coin with heads or tails, referring to a physical entity that could represent a bit in a state of (0), meaning heads, or (1), meaning tails. In qubits, we use the **Dirac** notation, which describes the state of a bit using "bra-ket":

|0> |1>

So, you can also imagine dealing with brackets that envelope 0 and 1. We already know that in superposition, the state of a particle can be 1 and 0 simultaneously, but when observed, the particle is forced to find a position. This phenomenon is called the **collapse of the wave function**. This phenomenon occurs when the waves (initially in a superposition state) reduce to a single state. This is caused by an interaction with the external world; for example, someone who spies on the channel of communication between the two particles.

When a pair of particles interact by themselves, their spin states can get entangled, which is what scientists call **quantum entanglement**.

When two electrons become entangled, the two electrons can only have opposite spins. This means that if one is measured to have an *up* spin, the second one immediately becomes a *down* spin:

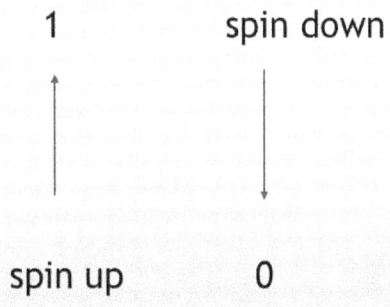

Figure 8.5 – The spin "up" and "down" of an electron

Even if we separate the two electrons so that they're arbitrarily far away and measure the spin of one, we will immediately know the spin of the second one. For example, if we measure an electron's spin in a California laboratory and know it's *up*, then we will know that the other one in the second laboratory in New Jersey will be *down*. It doesn't matter how far apart the two particles are.

So, we can say that the information traveled instantaneously and faster than the speed of light!

This theory had Einstein as a fierce opponent, who called this phenomenon *"spooky action at a distance."* But despite what Einstein said, entanglement is very useful for our technological world (as we will see very soon). Now, let's explore an experiment that can demonstrate that the entanglement of information is real and could be faster than light.

In 2015, a group of scientists led by Ronald Hanson, from Delft University in the Netherlands, set up an experiment to demonstrate that two particles that are entangled could communicate between themselves faster than the speed of light.

The experiment was carried out by setting two diamonds in two different laboratories, A and B, separated by a distance of 1,280 m. An electromagnetic impulse was shot to cause the emission from each point, A and B, of a photon in an entanglement state with an electron spin. The experiment demonstrated that when the two particles arrived simultaneously at a third destination, C, where a detector was installed, their entanglement was transferred to the electrons. We can see the experiment location mapped out in the following figure:

Figure 8.6 – Entanglement experiment (Delft University)

Recently, the group achieved an important technical improvement: the experimental setup is now always ready for *entanglement on-demand*. This means that the entanglement state between two particles can be obtained in the future on request, now allowing the development of quantum applications that were probably considered impossible earlier. If you are interested in learning more about this experiment, you can refer to the official Delft University website: `https://www.tudelft.nl/2018/tu-delft/delftse-wetenschappers-realiseren-als-eersten-on-demand-quantum-verstrengeling`.

Some examples of fields of action where this bizarre entanglement phenomenon is already used are as follows:

- Entangled clocks and all the applications behind the stock market and GPS
- An entangled microscope, built by Hokkaido University, to scrutinize microscopic elements
- Quantum teleportation, involving transporting information more than just matter
- Quantum biology for DNA and other clinical experimentations
- Our object of study: Q-Cryptography

So, it's just a combination of all these elements and properties to create a disruptive and fascinating application in **quantum key distribution** (**QKD**).

Before we look at this in more detail, let's talk about how this process came about and the elements that are combined in this process.

Q-Cryptography

Now that you have learned about the fundamentals of Q-Mechanics, we will talk about Q-Cryptography. The curious story of its first application began in the 70s, when a Ph.D. candidate, Stephen Wiesner, from Columbia University had an idea. He invented a special kind of money that (theoretically) couldn't be counterfeited: **quantum money**. Wiesner's quantum money mostly relied on quantum physics regarding photons.

Suppose that we have a group of photons traveling all in the same direction on a predetermined axis. Moving in space, a photon has a vibration known as the **polarization of the photon**. The following diagram shows what we are talking about:

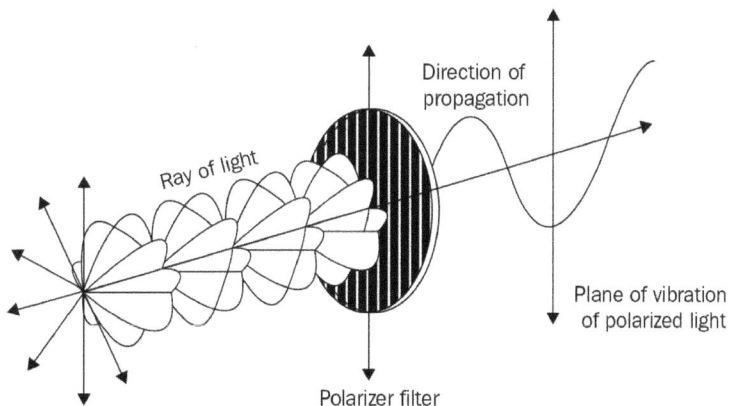

Figure 8.7 — Photon polarization

As you can see, photons spin their polarization in all directions. Still, if we place a filter, called a **polaroid**, oriented vertically, we will see that photons oriented vertically will pass across the filter 100% of the time. However, most of the other photons oriented in all directions will be blocked by the filter. In particular, the filter will block photons polarized perpendicularly to the filter 100% of the time, while photons oriented diagonally concerning the filter will pass (randomly) about 50% of the time. Moreover, these photons have to face a quantum dilemma: passing through the filter, they have to be observed, so they have to *decide* on their direction, *deciding* whether they want to assume a vertical orientation or a horizontal one.

This ability to block part of the photons is also explained in the experiment of polarized lenses. As shown in the following diagram, it's weird that if you add one more lens to the two already there, you can see much more light than before (see *Polarizer 3*):

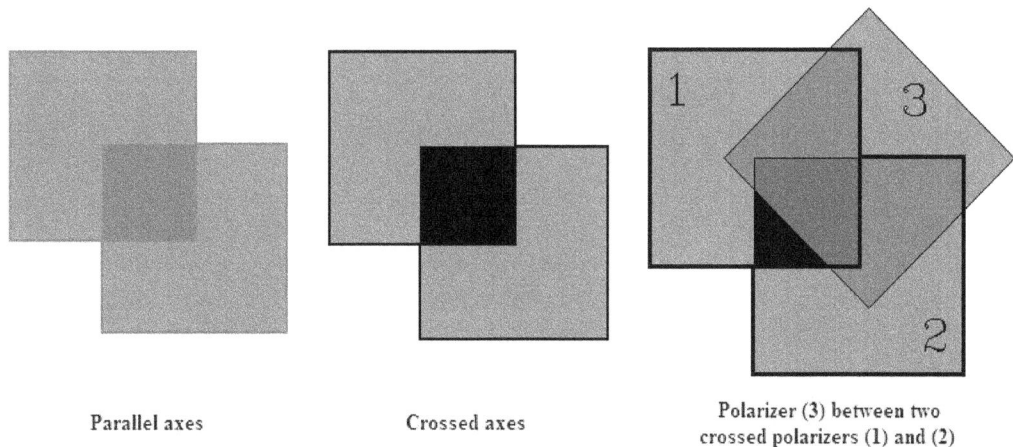

Parallel axes   Crossed axes   Polarizer (3) between two
crossed polarizers (1) and (2)

Figure 8.8 – The polarized filter experiment

The result of this experiment on the polarized filter shows the preculiarity of the Q-Mechanics world; otherwise, our logic would lead us to think that less light can pass through by adding one more lens and that it should ideally get darker.

So, what Wiesner proposed was to create a special banknote in which there are 20 light traps. These traps are oriented in four directions, as shown in the following diagram of a 1-dollar note. The filters can detect (by combining themselves with the serial number on the note) whether the money is real or fake:

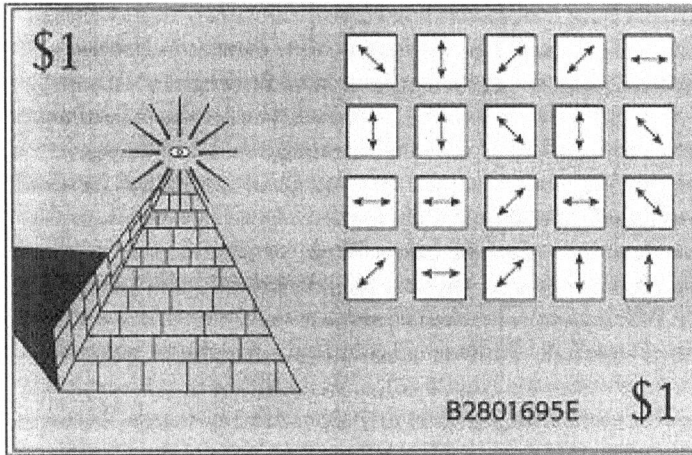

Figure 8.9 – Quantum 1-dollar banknote proposed by Wiesner

Only the bank that issues the notes can detect whether they are real or fake. In fact, due to the properties of Q-Mechanics, and in particular because of the indetermination principle, if someone wanted to counterfeit a note, it would be impossible to do so.

Let's see why an attacker would find it very hard to counterfeit one of these banknotes. In the preceding diagram, you can see the direction of the trap lights' orientation, but in reality, they are invisible so that nobody can see their orientation. Suppose Eve wants to copy the note; she can copy the serial number on the counterfeited note and try to figure out the traps' orientations.

If Eve decides to use a vertical filer to detect the polarization, she can detect all the vertical photons, and conversely, she will be sure that if the photon is horizontal, it will be blocked. But what about the others? Some of them will pass, but she will not be able to know exactly what polarization they have. So, the only thing Eve can do is randomly decide the polarization of all the photons so that they're different from the vertical and horizontal positions.

But now a problem occurs: if the bank runs a test on the banknote, they will know that it's fake. Because of the indetermination principle, only the bank knows the corresponding filter that's been selected for a determinate serial number.

Wiesner was always ignored when he talked about his invention of quantum money. It isn't a practicable invention to implement because the cost to set up such a banknote would be prohibitive; but this description introduces our next exploration well, which is the **quantum key exchange**.

Before we move on, just a note about quantum money: in his invention, Weisner intended to give banks the power to detect whether Eve had effectively attempted to counterfeit the note, but this concept is wrong. In fact, in the real world, the issue is not concerning banks (just a small part of the problem of counterfeit money affects banks). It is much more important that peer-to-peer users have control over judging the authenticity of banknotes. So, in my opinion, if quantum money ever comes out, it will be implemented in another way.

In the end, Wiesner found someone who listened to him, an old friend who was involved in different research projects and extremely curious about science: Charles Bennett. Charles spoke with Gilles Brassard, a researcher at the time in computer science at Montreal University; they started taking an interest in the problem and found out how to implement QKD, which we will explore next.

Quantum key distribution – BB84

Let's introduce BB84, the acronym that's Charles Bennett and Gilles Brassard developed in 1984 valid for QKD.

Now that we have learned about the properties of Q-Mechanics, we can use them to describe a technique for distributing bits (or better, **quantum bits**) through a quantum channel.

To describe what a quantum bit, also known as a **qubit**, is, we have to refer to the "quantum unit information" that's carried by qubits. Like traditional bits, `(0)` and `(1)`, qubits are mathematical entities subject to calculation and operations.

We will use a bi-dimensional vectorial complex space of unitary length to define a qubit. So, we can say that a qubit is a unitary vector acting inside of a two-dimensional space. So far, we can think of a qubit as a polarized photon, similar to the entanglement experiment we looked at in the previous section. I have already introduced the notation to represent the qubits inside "bra-ket": `|0>` and `|1>`.

I will not go too deep into the mathematics of Q-Mechanics, but we have to go a bit deeper to analyze some of the properties of qubits.

If an element carries some information (like qubits do), it is related to a form of computational process, which in this case is quantum computing. We will return to this concept later when we talk about quantum computers. To understand the geometrical representation of a qubit, take a look at the following diagram:

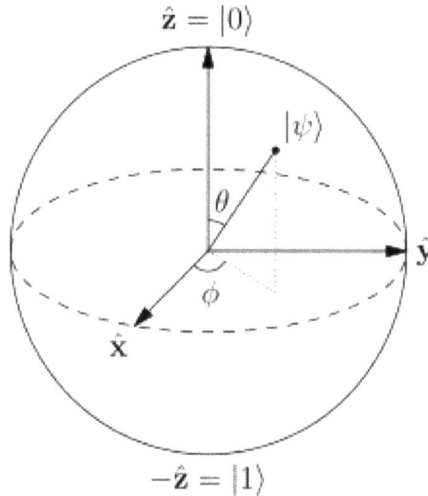

Figure 8.10 – Block sphere – the fundamental representation of a qubit

The difference between the classical computation we've used so far for all the classical bit operations, based on a Turing machine and Lambda calculus, and quantum computing is substantial. Quantum computing (related to computing with qubits) relies on superposition and entanglement properties. Qubits, as you saw when we discussed superposition, can be represented in a simultaneous state of (0) and (1), different from the normal bit, which can be only (0) or (1). We can mathematically represent qubits in the following way:

$$|0\rangle = \begin{pmatrix} 1 \\ 0 \end{pmatrix}$$

$$|1\rangle = \begin{pmatrix} 0 \\ 1 \end{pmatrix}$$

Figure 8.11 – Qubits represented in a mathematical way

Hence, considering a qubit as a unitary vector, it can be represented (as shown in the preceding diagram) as follows:

```
|ψ> = a |0> + b |1>
```

Here, $|a|$ and $|b|$ are complex numbers, so the following applies:

```
|a| + |b| = 1
```

In other words, we only know that the value of the sum is (1), but we never know how a photon will be in a superposition state, so the values of the $|a|$ and $|b|$ coefficients are unpredictable.

For our scope, that is enough for us to explain the algorithm behind a quantum distribution key.

Now that we've provided a brief introduction to the basic concepts of quantum calculus and qubit operations, we are ready to explore the quantum transmission key algorithm.

Step 1 – Initializing the quantum channel

Alice and Bob need a quantum channel and a classical channel to exchange a message. The quantum channel is a channel where it is possible to send photons through it. Its characteristic is that it must be protected from the external environment. Any correlation with the external world has to be avoided. As we have seen, an isolated environment is necessary for achieving the superposition and entanglement of a particle.

The classical channel is a normal internet or telephonic line where Alice and Bob can exchange messages in the public domain.

We suppose that Eve (the attacker) can eavesdrop on the classical channel and can send and detect photons, just like Alice and Bob can.

Since the interesting part of the system is just the quantum channel, where Alice and Bob exchange the key, we are going to observe what happens in this channel because this is the real innovation of this system. Then, we will make some considerations about transmitting the message through the classical channel.

Step 2 – Transmitting the photons

Alice starts to transmit a series of bits to Bob. These bits are codified using a base that's randomly chosen for each bit while following these rules.

There are two possible bases for each bit:

```
B1 = {|↑>, |→>}
```

and

```
B2 = {|↖>, |↗>}
```

If Alice choses B1, she encodes the following:

```
B1:     0 = |↑> and 1 = |→>
```

While if Alice chooses B2, then she will encode the following:

```
B2:     0 = |↖> and 1 = |↗>
```

Each time Alice transmits a photon, Bob chooses to randomly measure it with the B1 or B2 base. So, for each photon that's received from Alice, Bob will note the correspondent bit, (0) or (1), related to the base of the measurement that was used.

After Bob completes the measurements, he keeps them a secret. Then, he communicates with Alice through a classical channel and provides the bases (B1 or B2) that were used for measuring each photon, but not their polarization.

Step 3 – Determining the shared key

After Bob's communication, Alice responds, indicating the correct bases for measuring the photon's polarization. Alice and Bob only keep the bits that they have chosen the same bases for and discard all the others. Since there are only two possible bases, B1 and B2, Alice and Bob will choose the same base for about half of the bits that are transmitted. These bits can be used to formulate the shared key.

A numerical example is as follows.

Suppose that Alice wants to transmit the following bits sequence:

```
[0, 1, 1, 1, 0, 0, 1, 1]
```

Remember that the polarizations for each base are as follows:

```
B1:     0 = |↑> and 1 = |→>
B2:     0 = |↖> and 1 = |↗>
```

Alice randomly selects the bases:

- Alice's bases are B1, B2, B1, B1, B2, B2, B1, and B2.

- Alice's sequence is 0, 1, 1, 1, 0, 0, 1, 1.

- Alice's polarization is $|\uparrow\rangle, |\nearrow\rangle, |\rightarrow\rangle, |\rightarrow\rangle, |\nwarrow\rangle, |\nwarrow\rangle, |\rightarrow\rangle, |\nearrow\rangle$.

- Bob's choice is B2, B2, B2, B1, B2, B1, B1, and B2.

- Bob measurement is $|\nwarrow\rangle, |\nearrow\rangle, |\nwarrow\rangle, |\rightarrow\rangle, |\nwarrow\rangle, |\uparrow\rangle, |\rightarrow\rangle, |\nearrow\rangle$.

- The correct bases are   - $|\nearrow\rangle$,   - $|\rightarrow\rangle$, $|\nwarrow\rangle$, - $|\rightarrow\rangle, |\nearrow\rangle$.

- The corresponding bits are -   1   -   1,   0   -   1   1.

- So, the shared key is just the sequence of the selected bits [1, 1, 0, 1, 1].

The following is a representation of this scheme:

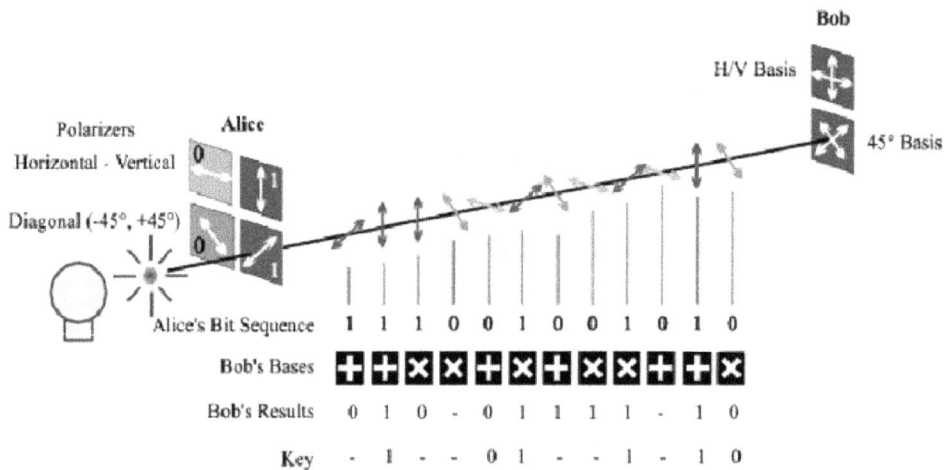

Figure 8.12 – Representation of QKD

At this point, we only need to check whether the two keys (from Alice and Bob) are identical.

Alice and Bob don't have to check the entire sequence but just a part of it to perform the parity bit control. If they check the entire parity sequence via the classical channel, it is obvious that Eve can spy on the conversation and steal the key. Therefore, they will decide to control only part of the sequence of bits.

Let's say that the key is of the order of 1,100 digits; they will check only 10% of the bits. After the parity control (made by a specific algorithm), they will discard the 100 bits that were processed in control and will keep all the others. This method lets them know that there were no errors in the transmission and detection process and whether Eve has attempted to observe the transmission in the quantum channel. The reason for this is related to Q-Mechanics' impossibility to observe a photon without forcing it to assume a direction of polarization (the indetermination principle). But if Eve measures the photons before Bob and allows the photons to continue for Bob's measurement, she will fail about half of the time. But when the same photon that's being measured by Eve arrives at Bob, he will have only a 25% probability of measuring an incorrect value. Hence, any attempt to intercept augments the error rating. That is why, after the third step, Alice and Bob performing parity control on the errors can also detect Eve's spying.

Now, the question of the millennium: *is it possible for Eve to recover the key?*

In other words, is this algorithm unbreakable?

Let's try to discover the answer in the next section, where we will analyze the possible attacks.

Analysis attack and technical issues

When we first described this algorithm, the assumption was that Eve has the same computational power as Bob and Alice and can spy on the conversations between them.

So, if Eve has a quantum computer in her hands and all the instruments to detect the photons' polarization, could she break the algorithm?

Let's see what Eve can do with her incommensurable computational power and the possibility of eavesdropping on the classical and quantum channels.

It looks like she is in the same situation as Bob (receiving photons). When Alice transmits the photons to Bob, even if the transmission comes through a quantum channel, if someone can capture the entire sequence of bits before Bob's selection, it's supposed that they will be able to collect the bits of the key. This could be true if Eve gets 100% of the measurements correct, but that is unthinkable for a key with a 1,000-bit length or more. Moreover, the attack will be discovered by Alice and Bob, as we will see now.

Let's analyze what can happen. Suppose that Alice transmits a photon with a polarization of $|\rightarrow\rangle$ and Bob uses a base (B1) that's the same as Alice's. If Eve measures the photon with the same base (B1), the measurement will not alter the polarization, and Bob will correctly measure the photon's state. But if Eve uses B2, then she will measure the two states, $|\nwarrow\rangle$, $|\nearrow\rangle$, with an equal probability. The photon that will arrive at Bob will only have half the probability of being detected as $|\rightarrow\rangle$ and correctly measured. Considering the two possible choices Eve has to determine a correct or incorrect measurement, Bob will only have a 25% chance of measuring a correct value.

In other words, each time Eve attempts to measure augments the probability of an error rating occurring in Alice and Bob's communication. Therefore, when Alice and Bob go to check on the parity bits, they will identify the error and understand that an attack is being attempted. In this case, Alice and Bob will repeat the procedure, (hopefully) avoiding the attack again.

While analyzing the system, I endorse the thesis that this system is theoretically invulnerable, but I also agree that it is only a partial solution, mainly because this is not a quantum transmission message, but a quantum exchanging key. The reason I said it's theoretically invulnerable is that Eve can be detected by Bob and Alice by relying on the indetermination principle of Heisenberg. However, a **man-in-the-middle attack**, even if it doesn't arrive to recover the key (Eve can attempt to substitute herself with Bob), can block the key being exchanged between Alice and Bob, inducing them to repeat the operation infinite times.

In a document from the **NSA**, the **National Security Agency** of the United States, you can read about some weaknesses related to QKD and why it is discouraged that you implement it: `https://www.nsa.gov/Cybersecurity/Quantum-Key-Distribution-QKD-and-Quantum-Cryptography-QC/`.

What is to be understood is whether the NSA's concerns are related to the real possibility of implementing such a system or the possibility that hackers and other bad people could take advantage of this strong system.

In the NSA's document, we note some issues related to the implementation of QKD in a real environment. The NSA's notes are explained as follows:

- **QKD is only a partial solution**: The assumption is that the protocol doesn't provide authentication between the two actors. Therefore, it should need **asymmetric encryption** to identify Alice and Bob (that is, a **digital signature**) by canceling quantum security benefits. This problem could be avoided by using a quantum algorithm of authentication for implementing a quantum digital signature, if possible. Unfortunately, with the current state of the art, the way of identifying through a quantum algorithm and the non-repudiation of the message gives only a probabilistic result, far from the deterministic digital signature of the classical algorithms. Another solution is using a MAC (a function for authenticating symmetric algorithms), but again, it may *not* be quantum-resistant in the future.

- **Costs and special equipment**: This problem is concerned with higher costs and implementation problems for users. Moreover, the system is difficult to integrate with the existing network equipment.

- **Security and validation**: The issues, in this case, are related to the hardware being adapted to implement a quantum system. The specific and sophisticated hardware that's used for QKD can be subject to vulnerabilities.

- **Risk of denial of service**: This is probably the biggest concern of QKD. As I mentioned previously, Eve can repeat the attack on the quantum channel, sequentially causing Alice and Bob to never be able to determine a shared key.

As we have seen, the QKD method is not unerring. There is a remote possibility that someone, in practice, can break the system.

> Note
> QKD is just a part of the cryptographic transmission algorithm; the second part of the system is the encryption algorithm that's used to send the message, [M].

This is a fundamental part of Q-Cryptography's evolution because to be complete, the system should guarantee the transmission of the message, [M], through a quantum channel, but that needs another quantum algorithm. Moreover, the quantum authentication protocol remains unsolved.

We can say that QKD could be compared to **Diffie–Hellmann**, as **RSA** could be compared to a **quantum message system transmission**.

There are various choices we can make when it comes to transmitting messages in classical cryptography; we can use **Vernam** (as we saw in *Chapter 1*, *Deep Diving into the Cryptography Landscape*), which, combined with a random key (QKD), is theoretically secure, or another symmetric algorithm. But again, we don't rely on a fully quantum-based encryption system, which is autonomous and independent of classical mathematic problems.

Returning to the QKD algorithm, we have to find out why this algorithm is theoretically impossible to break, even with a quantum computer.

If an attacker decrypted a key generated with Q-Cryptography, the result of this action would probably be devastating for the entirety of Q-Mechanics. This is because one of the axioms that Q-Mechanics bases its strength on will be inconsistent: we are talking about the indetermination principle. If something like this happens, then the complete structure of Q-Mechanics would probably be involved in a revision process that would shock the pillars of the theory and all its corresponding physical and philosophical implications.

To establish this concept, I propose a theory: suppose that, in the future, someone will discover that the photons passing between the point of Alice's transmission and the point of Bob's measurement are truly random, but they follow a pre-determinate scheme. Suppose that this scheme is π or a Fibonacci series. In this case, the bits are effectively random, but their randomicity will be known so that Eve can forecast what the polarized scheme to adopt is.

What happens, for example, if we recognize that a sequence of digits such as "1415" is included in π?

As shown in the following code, π is a sequence composed of an infinite number of digits that follow the number 3, starting with "1415".... after the decimal point:

```
π = 3.14159265358979323846264338327950288419716939937510582097
4944592307816406286208998628034825342117067982148086513282306
64709384460955058223172535940812848111745028410270193852110555
96446229489549303819644288109756659334461284756482337867831652
7120190...
```

We can go on following the π sequence and recovering all the other digits after 1415 that are: 92653589...

Undoubtedly, I agree with the NSA's report when they said that QKD is an incomplete algorithm. It misses the second part of the algorithm, which completes the encryption of the message: [M] into the quantum channel. By using BB84, this stage is devolved to classical cryptography because after exchanging the key (in the quantum channel), we must encrypt the message using a classic encryption algorithm.

As I have always been against hybrid schemes, I have projected that a quantum encryption algorithm would be valid for quantum message system transmission. Its name is **QMB21**, and it will hopefully become a standard for the next Q-Cryptography transmissions. Due to issues linked to the process of patenting, it is not possible to explain the algorithm here. I hope to be able to update this section with more details on this innovation soon.

Now that we have analyzed the Q-Cryptography key distribution, another challenge awaits: exploring quantum computing and what it will be able to do when its power increases.

Quantum computing

In this section, I will talk about **quantum computing** and **quantum computers**. The difference between the two is that quantum computing is the computational power that's expressed by a quantum system, while a quantum computer is the system's physical implementation, which is composed of hardware and software. The first idea of a quantum computer was originally proposed by Richard Feynman in 1982; then, in 1985, David Deutsch formulated the first theoretical model of it. This field had a big evolution in the last decades; some private companies began to take action and experiment with new models of quantum computers. Some of them, such as **D-Waves** and **Righetti Computing**, raised millions of dollars for the research and development of these machines and their relative software:

Figure 8.13 – D-Waves quantum computer

As you should already know, normal computers work with bits, while quantum computers work with qubits. As we have also already seen, a qubit has a particular characteristic in that it can be simultaneously in the state of (0) and (1). Just this information by itself suggests a greater capability of computation concerning a normal computer.

A normal computer works by performing each operation one by one until it reaches the result. Conversely, a quantum computer can perform more operations at the same time. Moreover, many qubits can be linked together in a state of entanglement. This phenomenon creates a new state of superposition over the multiple combinations of the entangled qubits, elevating the grade of computation of a quantum computer exponentially compared to a normal one. So, we are not only talking about a parallelization process of computation but also an exponential computation process when augmenting the number of qubits in the game. Therefore, to give you an idea of the computational power that's created by a quantum computer, n qubits can generate a capability to elaborate information 2^n times compared to classical bits.

Let's consider the preceding sentence. In a state of superposition or entanglement, 8 qubits, for instance, can represent any number between 0 and 255 simultaneously. This is because $2^8 = 256$, so we have 256 possible configurations each time, which, expressed in binary notation, looks like this:

```
0 = (0)2 ———-> 255 = (11111111)2
```

We can understand the supremacy of a quantum computer, especially if we deal with two specific problems of extreme interest to cryptography: the factorization problem and searching on big data.

These two arguments are considered very hard to solve for a classical computer; we have seen, for instance, the RSA algorithm and other cryptographic algorithms relying on factorization to generate a one-way function that's able to determine a secured cryptogram. However, the community's recurring question is, *how long will classical cryptography resist when a quantum computer reaches enough qubits to perform operations that are considered unthinkable?*

We will try to answer this question after exploring the potential of a quantum computer and seeing what we can do with one by projecting one adapted for our scopes.

As we mentioned previously, a quantum computer can simultaneously perform all the state bases of a linear combination. Effectively, in this sense, a quantum computer can be considered an enormous parallel machine that's able to calculate huge states of combinations in parallel.

For example, to understand what a quantum computer can do, let's take three particles: the first of orientation 1 and the last two with an orientation of 0 (relative to some base). We can represent this as follows:

```
|100>
```

The quantum computer can take |100> as input and give some output, but at the same time, it could take a *normalized combination* of the three particles (of length = 1) with a state base as input, as follows:

```
1/√3 (|100>  +  |100>  +  |100>)
```

As a result, it gives an output in only one operation as it computes only the state base.

After all, the quantum computer doesn't know whether a photon is in one state base or multiple combinations of states until a measurement occurs (the indetermination principle).

This ability to elaborate combinations of linear states simultaneously determines the quantum supremacy of quantum computers.

So, while a classical computer (even the most powerful supercomputer on the earth) requires a string of bits as input and returns a string of bits as output (even if it's parallelizing billions of operations), it will not be able to perform a function that accepts the following sum as input:

```
1/C ∑ |x>
```

Here, C is a normalization factor of all the possible states of the input. The function would return the following as output:

```
1/C ∑ |x,f(x)>
```

Here, |x,f(x)> is a succession of bits longer than |x|, which represents both |x| and f(x).

To understand this concept, you can think of having a sum of bits, (C), and in this sum, all the possible states of (m) qubits in a superposition are hidden. So, we can say that (C) represents the number of all the states that the qubits are in at the same time.

You can imagine how fantastic that could be! But still, there is a problem: the measurement. We have seen that when we measure the quantum computation, this measurement *freezes* the qubits in only one state, which is governed by randomness. Therefore, we can force the system *to appear* in a state, as follows:

```
|x0,f(x0)>
```

This is for some x0 value that's randomly chosen, so all the other states are going to be destroyed.

In this case, the ability to program a quantum computer makes it possible to create a quantum computation that provides a very high-probability result of getting the desired output.

To understand the process of implementing a quantum algorithm, let's look at a candidate algorithm that could wipe out a great part of classical cryptography: Shor's algorithm.

Shor's algorithm

In a recent interview with David Deutsch (considered one of the fathers of quantum computing), the interviewer posed a question about the philosophy of superposition, asking David, *"In what way does the Quantum Computing community give credit to the hypothesis of a multiverse?"*

David's answer refers to Q-Cryptography and, in particular, he tries to demonstrate the existence of the multiverse through the factorization problem. I will try to paraphrase David's answer in the following paragraph:

Imagine deciding to factorize an integer number of 10,000 digits, a product of two big prime numbers. No classical computer can express this number as the product of its prime factors. Even if we take all the matter contained in the universe and transform it into a supercomputer, which starts to work for a time long, such as the universe's time, this instrument will not be able to scratch the surface of the factorization problem. Instead, a quantum computer should be able to solve this problem in a few minutes or seconds. How is this possible?

Whoever is not a solipsist has to admit that the answer is linked to some physics process. We know that not enough power exists in the universe to calculate and obtain an answer, so something has to happen compared to what we can see directly. At this point, we have already accepted the structure of the multiverse (otherwise known as superposition).

The way (described by Deutsch) in which a quantum computer works is as follows:

"The universe has evolved into multiple universes and in each, different sub-calculation is performed. The number of sub-calculations is enormously bigger than all the atoms contained in the known universe. In the end, the results became englobed to obtain a unique answer. Whoever neglects the existence of multiverses has to explain how the factorization process works."

After this interesting digression into *quantum computational power*, let's dive deeper and analyze Shor's algorithm mathematically and logically; this algorithm is the true representation of what happens in a quantum computer at work.

The **hypothesis**, by Fermat's Last Theorem, is as follows: say you find two (non-trivial) values, (a) and (r), so the following applies:

```
a^r ≡ 1 (mod n)
```

Here, there is a good possibility of factorizing (n).

We start to randomly choose a number, (a), and consider the following sequence of numbers:

```
1, a, a^2, a^3, … , a^n (mod n)
```

The **thesis** is that we can find a period for this sequence that we call (r), which will be repeated every (r) time when we elevate (a) to (r) so that we will have solved the following equation for (r), which always gives (1) as a result:

```
a^r ≡ 1 (mod n)
```

Therefore, we want to program our quantum computer to find an output, (r), to be able to factorize (n) *with a high probability*.

Step 1 – Initializing the qubits

We choose a number, (m), so that the following applies:

```
n^2 ≤ 2^m < 2^n
```

Supposing we choose m = 9, this means that we have 9 bits initialized with their state all in (0), and we mathematically represent it as follows:

```
m = 9 : |000000000>
```

Changing the axes of the bits, we can transform the first bit into a linear combination of $|0>$ and $|1>$, which gives us the following:

```
1/√ 2  (|000000000> +  |100000000>)
```

By performing a similar transformation on every single bit of (m) inside *bra-ket* until the *m-bit to obtain the quantum state*, we get a sequence like this:

```
1/√ 2^m (|000000000> +  |000000001>+  |000000010> + . . . +
|111111111>)
```

Remember that m = 9, so 2^m = 512; we will have 512 multiple combinations of bits, given by the first, $|000000000$ = 0, and the last, $|2^m-1> = |111111111> = 511$ (in decimal notation).

In a sum like this, all the possible states of the m qubits are overlying in a superposition. Now, for simplicity regarding notation, we will rewrite the m qubits in cardinal numbers:

```
1/√ 2^m (|0> +  |1>+  |2>+ . . . + |2^m-1>)
```

The preceding functions are just the same as the previous ones but are represented in cardinal numbers.

Step 2 – Choosing the random number, a

Now, we must choose a random number (a) so that the following applies:

```
1 < a < n
```

Quantum computing computes the following function:

```
fx ≡ a^x (mod n)
```

It returns the following:

```
1/√ m^2 (|0, a^0 (mod n)> + |1, a^1 (mod n)>+ |2, a^2 (mod n)>+
 . . . + |2^m-1, a^2m-1 (mod n)>)
```

When visualizing this function, you can lose the sense of representation because it appears rather complex; so, to simplify it, forget the modulus, n, that's repeated for each operation and rewrite the previous functions like so:

```
1/√ m^2 (|0, a^0 > + |1, a^1 >+ |2, a^2>+ . . . + |2^m-1, a^2m-
1>)
```

Again, consider that all the operations are in (mod n), but focus on the sense of the functions: it says that we have to elevate (a) to all the (2^m-1) states of the m qubits.

However, so far, the *state* of this representation doesn't provide any more information compared to a classical computer.

Now, it's time to look at an example to help you understand what we are doing.

Suppose that n = 21 (a very easy number to factorize); however, the size of the number is not important now but the method is.

Here, we choose randomly a = 11.

So, we calculate the values of 11^x (mod 21) that have been transposed into the function:

```
1/√ 512 (|0, 1> + |1, 11> + |2, 16> + |3, 8> + |4, 4> + |5, 2>
+ |6, 1> + |7, 11> +
|8, 16> + |9, 8> + |10, 4> + |11, 2> + |12, 1> + |13, 11> +
|14, 16> + |15, 8> +
|16, 4> + |17, 2>. . . . . . . . . . . .+ |508, 4> + |509, 2> +
|510, 1> + |511, 11>)
```

As you can see, this is a development of the entire sequence, where 1/√ 512 = 1/ C and C is the number of states of m = 2^9.

Each qubit is composed of a sequence number, (a^x (mod 21)). Take the following example:

```
|1, 11> ≡ 11^1  (mod 21) = 11
```

Step 3 – Quantum measurement

If we measure only the second part of the sequence, $a\char`^x$, then the function collapses in a state that we cannot control. However, this measurement is the essence of the system. In fact, because of this measurement, the entire sequence collapses for some random base, $(x0)$, so the following applies:

```
|x0, a^x0>
```

Now, it should be clearer to you that we can force the system *to appear* in a state, as follows:

```
|x0, f(x0)>
```

So, after the measurement, the state of the system is fixed (in other words, a number appears) in only one base $(x0)$ that we suppose to be as follows:

```
x0 = 2
```

Forgetting all the other values of $a\char`^x$ (that we have previously calculated) and picking up only the ones that have a value equal to 2, we will extract the following highlighted values:

```
1/√ 512 (|0, 1> + |1, 11> + |2, 16> + |3, 8> + |4, 4> + |5, 2>
+ |6, 1> + |7, 11> +
|8, 16> + |9, 8> + |10, 4> + |11, 2> + |12, 1> + |13, 11> +
|14, 16> + |15, 8> +
|16, 4> + |17, 2>. . . . . . . . . . . .+ |508, 4> + |509, 2> +
|510, 1> + |511, 11>)
```

To simplify the notation, we must discard the second part of the qubit (2) because it is superfluous:

```
1/√ 85 (|5> + |11> + |17>. . . . . . . .+ |509>)
```

Here, 85 stands for the number of m qubits selected.

Now, if we make a measurement, we will find a value, x, so that the following applies:

```
11^x ≡ 2 (mod 21)
```

Unfortunately, this information is not useful for us. But don't lose hope.

Step 4 – Finding the right candidate, (r)

Recall that we have programmed our quantum computer with a specific scope: to find the output in the system that can factorize (n). To reach our goal, we have to find the value of (r) so that the following applies:

```
a^r ≡ 1 (mod n)
```

Generally, when we deal with functions such as a^x (mod n), we know that the values of (x) are periodic, with a period of (r) where $a^r = 1$ (mod n).

So, recalling the previous function in which the selected elements appear, we have the following:

```
|5> + |11> + |17>. . .  . . . .+ |509>
```

As you can see, the period is $r = 6$.

Therefore, we can verify whether the following applies:

```
11^r ≡ 1 (mod 21)
11^6 ≡ 1 (mod 21)
```

Here we go!

Hence, $r = 6$ is a very good candidate for us.

Note

In this example, we can see that $r = 6$ is the period of the function, though it's not always easy to work out. There is a mathematical way to discover the period of a function. Using the **Quantum Fourier Transform (QFT)**, we can discover the period (if it exists). If you want to continue looking at the demonstration of Shor's algorithm, you can skip the next section and go directly to the last step of Shor's, coming back to QFT on your second read.

Quantum Fourier Transform

Let's discover a little bit more about the **Fourier Transform** (**FT**) and QFT.

FT is a mathematical function. It can be intuitively thought of like a musical chord in terms of volume and frequency. The FT can transform an original function into another function, representing the amount of frequency present in the original function. The FT depends on the spatial or temporal frequency and is referred to as a time domain. It is represented by a graphic that shows the frequency that was detected, as shown in the following diagram:

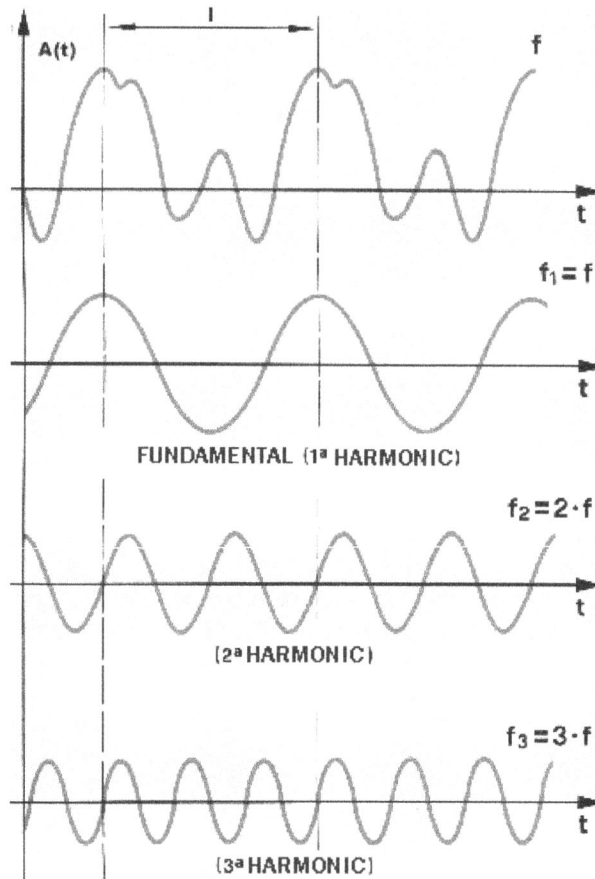

Figure 8.14 – Fourier Transform

To understand how this works, let's look at an example of FT taking an arithmetic succession of numbers, like so:

```
1, 3, 7, 2, 1, 3, 7, 2
```

As you can see, we have the first four numbers – 1, 3, 7, 2 – which get repeated two times. In other words, we can break the succession into two parts:

```
1,3,7,2 | 1,3,7,2
```

> **Note**
> The | symbol represents, in this case, splitting the succession into two parts.

Here, we can say that this succession has a length of 8 and a period of 4. The length, divided by the period, gives us the frequency (`f`):

```
f = 8/4 = 2
```

In this case, `f = 2` is the number of times that the succession is repeated.

Now, we can enounce the general rule of FT. Suppose we have a succession that's 2^m in length for any integer, `m`:

```
[a0, a1, ...., a2^m-1]
```

We can define the FT with the following formula:

$$FT\ (X) = 1/\sqrt{2}\char94 m \sum_{c=0}^{2\char94 m-1} e\char94 (2\pi icx/2\char94 m)\ ac$$

We can define the following part of the formula with `j`:

```
j = e ^ (2πicx/2^m)
```

The `x` parameter runs in the following range:

```
0 ≤ x < 2^m
```

With the FT formula, you can find the frequency of the succession while using determinate parameters as input. If we suppose, for example, `cx = 1` and `m = 3`, then we have the following substitution:

```
j = e^(2πi/8)
```

By substituting the different values of x in the equation with `j`, we obtain the following result by exploding the formula for `FT(1)`:

```
√ 8 FT(1) = 1+ 3j + 7j^2 + 2j^3 + j^4 + 3j^5 + 7j^6 + 2j^7 = 0
```

Since all the terms of the succession get eliminated, we have the following:

- `FT(X0) with x = 0 ---> e^0 = 1`, which is the result of `FT(X0) = 1`
- `FT(X4) with x = 4 ---> e^πi = -1`, which is the result of `FT(X4) = -1`

So long as `j^4 = -1`, all the terms will be deleted and we will obtain `FT(1) = 0`.

Instead, for `FT(0)`, the result of the numerator addition corresponds to the sum of all the terms of the sequence:

```
1 + 3 + 7... + 2 = 26
```

Continuing, we have the following:

```
FT(0) = 26/√ 8              FT(2) = (-12 +2i) /√ 8
FT(4) = 6/√ 8              FT(6) = (-12 -2i) /√ 8
```

For all the other terms, `FT(1) = FT(3) = FT(5) = FT(7) = 0`.

In other words, the FT confirms what we have deducted since the beginning of this example – that is, in the succession 1, 3, 7, 2, 1, 3, 7, 2, the peaks of the function correspond with the non-zero values of (FT) that are multiples of 2: the frequency.

This result will be very useful in Shor's algorithm to discover the period of the function, as we will see shortly.

But we want to program our algorithm in a quantum computer, so we need a version of FT that's been adapted for quantum algorithms.

QFT is what we need to detect the frequencies that we use to find the period (r) in Shor's algorithm. It is defined on a base state of $|X\rangle$ with $(\text{or} \leq x < 2^m)$:

$$QFT\,(|X\rangle) = 1/\sqrt{2^m} \sum_{c=0}^{2^m-1} e^{\wedge}(2\pi i c x/2^m)\ |c\rangle$$

As you can see, the FT (shown previously in this chapter) and QFT (shown here) are very similar. The difference is that QFT is acting in a quantum environment, so the c parameter of FT now appears in a quantum state, $|c\rangle$, and the same is valid for $|X\rangle$.

Returning to our quantistic example of Shor's algorithm, we can apply the QFT to it to discover the frequency:

```
QFT (1/√ 85 (|5> + |11> + |17>. . . . . . . .+ |509>))
```

We obtain the following sum:

```
1/√85 ∑ g(c) |c)
```

For `c`, this runs from 0 to 511. Here:

$$g(c) = 1/\sqrt{512} \sum_{0 \leq x \leq 512} e^{\wedge}(2\pi i c x/512)$$

This is the FT of the following succession:

```
0,0,0,0,0,1,0,0,0,0,0,1..... 0,0,0,0,0,1,0,0
```

By applying a break symbol to the succession, we can visualize the frequency:

```
0,0,0,0,0,1 |0,0,0,0,0,1 |..... 0,0,0,0,0,1 |,0,0
```

As you can see, the frequency of this succession is around `512/6 ~ 85`.

In this sinusoidal function, which is given by waves in which there are high and low points, the picks (highest points of interest) of the function correspond with the following:

```
c = 0, 85, 171, 256, 314, 427
```

We expect that `427` is a multiple of the frequency we are looking for:

```
427 ~ j f0
```

This applies so long as the following is true:

```
427/512 ~ 5/6
```

So, we expect that `r = 6` is the period.

> **Note**
>
> I didn't show all the mathematical passages that bring us to the result of `r = 6`, but the intention is to give you a demo of the possibility to apply QFT to this algorithm. For a deeper analysis of QFT, go to the following learn session of the IBM Quantum Lab project: `https://qiskit.org/textbook/ch-algorithms/quantum-fourier-transform.html`.

Now that we know what QFT does and how to use it in Shor's algorithm, let's perform factorization through a quantum algorithm.

Step 5 – Factorizing (n)

The last step of Shor's algorithm uses the **greatest common divisor (GCD)** to calculate whether the candidate (r) gives back the two factors of (n).

As you probably remember from high school, the GCD is the greatest positive common divisor between two numbers. It can also be used to reduce a fraction to its smallest possible divisors.

Regarding the Mathematica software, there's a very useful reduction function that is implemented when you divide by two integers.

In our case, if r = 6 and n = 21, we just divide the two numbers:

```
In[01]: = GCD[21, 6]
Out[01]: = 3
```

Here, I have told Mathematica to perform GCD[21,6], finding the number (3): the GCD between (21) and (6).

Now, by reducing the fraction (the operation consists of dividing both 21 and 6 by 3), we get 7 as the second output:

```
In[02]: = 21/6
Out[02]: = 7/2
```

Therefore, 21 = 7*3.

We got the solution!

Notes on Shor's algorithm

Peter Shor presented this algorithm when the quantum computer hadn't been experimented with and built up in practice. Here, I provided an example of how it could be possible to perform a factorization in a theoretical way on a quantum computer, but at the moment, no machine exists that can perform this algorithm in a scalable way.

Moreover, Shor's algorithm is not deterministic. Still, it gives a probabilistic grade to find the factors, even if, generally, after a few steps (if the first candidate that's found is discarded), it will be possible to find a good candidate for the factorization of (n).

From the opposite point of view, if a scalable quantum computer were to be implemented and ready to run this algorithm in the future, it will probably be the end of classical cryptography.

But now, it's time to answer our initial question about quantum computers, reformulated appropriately: what are the consequences of cryptography when the quantum computer reaches enough qubits to perform Shor's algorithm?

Let's discuss this topic in the final section of this chapter, dedicated to **post-Q-Cryptography**.

Post-Q-Cryptography

It is useful to point out that post-Q-Cryptography has nothing to do with Q-Cryptography, except for the fact that it could be considered resistant to quantum computing (maybe it is only a chimera). When we talk about post-Q-Cryptography, we refer to classical algorithm candidates being resistant to quantum computing.

If quantum computing were to raise its power, expressed in a satisfying number of qubits, and were implemented on appropriate hardware able to perform the quantum computation, many of the algorithms that have been discussed in this book would be breached. However, nowadays, a quantum computer occupies an entire room of space and works through hardware made by several components (most of them prepared to keep the qubits in a state of entanglement at frozen temperatures). As soon as the dimensions of the hardware are reduced and its calculation power increases, most of the following algorithms will probably fail in terms of timing, which could swing between 1 hour and a few days: RSA and Diffie–Hellman, Schnorr protocols, and most of the zero-knowledge protocols – even elliptic curves; it is supposed that most elliptic curves implementations will be breakable.

Based on a document provided by the **European Commission for Cybersecurity (ENISA)** released in February 2021, some categories of algorithms could support the shock wave of quantum power computation.

In this study, the commission commented on the advent of quantum computing:

> *"It is thus important to have replacements in place well in advance. What makes matters worse is that any encrypted communication intercepted today can be decrypted by the attacker as soon as he has access to a large quantum computer, whether in 5, 10, or 20 years from now; an attack known as retrospective decryption."*

The categories of algorithms that are reputed to be resilient to quantum computing, many of them previously evaluated by NIST, are as follows:

- **AES**: This is considered a good post-quantum candidate: we saw this algorithm in *Chapter 3, Asymmetric Encryption*.

- **Hash functions**: We saw these functions in *Chapter 4, Introducing Hash Functions and Digital Signatures*.

- **Code-based cryptography**: This uses the technique of error-correction code, based on a proposal by *McEliece*.

- **Isogeny-based cryptography**: This is a kind of cryptography based on adding points to elliptic curves over finite fields.

- **Lattice-based cryptography**: Among these algorithms, the NTRU algorithm is probably one of the best candidates that resists quantum computing.

All these categories are supposed to be quantum-resistant, based on evaluations given on mathematics and cryptoanalysis of the mentioned systems. However, some skeptical thoughts are spreading among the community, most of which are often calmed down by announcements about minimizing problems.

Summary

In this chapter, we analyzed Q-Cryptography. After introducing the basic principles of Q-Mechanics and the fundamental bases of this branch of science, we deep-dived into QKD.

After that, we explored the potential of quantum computing and Shor's algorithm. This algorithm is a candidate for wiping out a great part of classical cryptography when it comes to implementing a quantum computer that will get enough qubits to boost its power against the factorization problem.

Finally, we saw the candidate algorithms for the so-called post-Q-Cryptography; among them, we saw AES, SHA (hash), and NTRU.

With that, you have learned what Q-Cryptography is and how it is implemented. You have also learned what quantum computing is and, in particular, became familiar with a very disruptive algorithm that you can implement in a quantum machine: Shor's algorithm.

Finally, you learned which algorithms are quantum-resistant to the power of the quantum computer and why.

These topics and the concepts that were explained in this chapter are essential because the future of cryptography will rely on them.

Now that you have learned about the fundamentals of Q-Cryptography and most of the principal cryptography algorithms that have been produced until now, it is time to move on to the next chapter, where we will look at the Crypto Search Engine. The next chapter will provide you with knowledge about search engine logic. Then, we will explore an innovative search engine platform that works with encrypted data, developed by my team and me.

Section 4: Homomorphic Encryption and the Crypto Search Engine

This section shows the theoretical basis of homomorphic encryption and isomorphic search, the math behind search engines, and the practical implementation of the cryptography presented in this book.

This section of the book comprises the following chapter:

- *Chapter 9, Crypto Search Engine*

9
Crypto Search Engine

This chapter will discuss the technique of searching in encrypted data and provide some basic information on search engines in general and the math behind them. Then we will discover an innovative search engine that is able to work with encrypted data: the **Crypto Search Engine** (CSE).

This invention was projected and developed by Cryptolab, `https://www.cryptolab.us/`, a cybersecurity company active in data privacy and protection, founded in 2012 as a cryptography laboratory by me and my co-founders, two data science engineers, Alessandro Passerini and Tiziana Landi. The aim of Cryptolab is to give more privacy and security to the data transmitted and processed over the internet and then stored and shared in the cloud. Another website you can refer to is the web page of CSE: `https://www.cryptolab.cloud/`.

We are going to talk about homomorphic encryption and, in particular, homomorphic searching, which is one of the most fascinating problems I have faced during my career. I am excited to explain what inspired me to start this work and its genesis; moreover, you will discover the math behind search engines and their relative algorithms.

So, in this chapter, we will cover the following topics:

- Introduction to CSE – homomorphism
- The math and logic behind search engines
- Introduction to code theory and graph theory
- CSE explained
- Applications and possible evolutions of CSE

Let's take a deep dive into these topics to learn the secrets behind homomorphic search.

Introduction to CSE – homomorphism

The genesis of CSE dates back to 2014 when I was struggling for several months with a new method of *factorization*. You can understand that the factorization problem and *search in blind* are strictly related to each other. Both these problems, factorization and searching among big data, have similar complexity.

Moreover, both these problems have their domain inside *P=NP*, meaning some problems are *easy* to solve (*P*) while others are *very hard* to solve (*NP*) even if they are supposed to get a very high or infinite level of computation.

Most scientists and data science engineers are convinced that *P≠NP* or all *NP* problems are intrinsically complex and cannot be solved with a polynomial algorithm. I don't think so; I am more interested in finding *solutions* to complicated problems as opposed to saying that solutions don't exist. I am also convinced that there are different ways to obtain a solution. For example, Fermat's Last Theorem has been solved by Andrew Wiles, as was proven through a demonstration of about 100 pages (take a look at this paper for a formal demonstration, `https://staff.fnwi.uva.nl/a.l.kret/Galoistheorie/wiles.pdf`). However, I am convinced that it could be solved in just a few pages and I have some ideas about it. Likewise, suppose we use an efficient quantum processor to perform Shor's algorithm, as we saw in *Chapter 8, Quantum Cryptography*. In this case, we can obtain the solution to the factorization of big semiprimes in a few seconds. In the same way, if we use a quantum algorithm such as Grove's algorithm to process a search inside a huge database composed of billions and billions of instances (theoretically), we can find the solution quickly.

Coming back to the story of CSE, after we completed **version 1.0** of our *cryptographic library* in 2014 at Cryptolab, I started to implement an algorithm that could search for data (in blind). The Crypto library project consisted of amalgamating into one *Crypto-Library* all the inventions developed by me and my team of researchers between 2009 and 2014. We called this project *Cryptoon* and it turned out to be a C++ library consisting of not only our cryptography algorithms but also symmetric/asymmetric, zero-knowledge, and authentication algorithms, most of which I have already discussed in this book.

At the time, I was enthusiastic about the discoveries made in the field of *homomorphic encryption*. This kind of encryption has a peculiarity: it performs the manipulation of data (operations on it) in an encrypted way and gives back the encrypted mode results. After the results have been achieved, those who detect the decryption key can retrieve the encrypted result by decrypting it and obtaining the plaintext result. Let's make a scheme to better understand this complicated logic:

The operations on encrypted data generate a consequent encrypted result E [Z] as follows:

```
Plaintext (A),(B) --------> Encrypted Data E[A], E[B] -------->
Operations on Encrypted Data = Encrypted result:

E[A]∀o E[B] = E[Z]
```

The decryption of E [Z] results as follows:

```
E[Z] --------> Decryption of Encrypted Result [Z] is:
D(Z) = (A) ∀o(B) = (Z) (Plaintext).
```

> **Note**
>
> I adopted a symbol of my invention: ∀o to denote any mathematical and logical operation (or set of operations) performed between (A) and (B). ∀o stands for ∀ = for all; o = operations.

But don't worry. We will have a better understanding of these concepts later in this chapter. So, let's now introduce the concept of homomorphism.

Homomorphisms are operations performed on *isomorphic elements*. We can find isomorphisms in nature, such as in mathematics. In nature, for example, we can see that the backs of the wings of some butterflies reproduce a concentric pattern in the shape of wings. The skeleton of a person relates to his body, or his shadow, which is a form of isomorphism: elements belonging to a person, an animal, or something else generated, related, or derived from itself.

However, even though we cannot distinguish the face of a person by looking at their shadow, we can certainly recognize some elements of them, such as their shape or the curly hair, that identify this individual. We are not talking about similarity, but a relational property between two elements constituting part of the same thing or generated by the same element. In the following figure, you can see the difference between the two photos: the left one represents isomorphism because the shadow is the reflection of the body on the sand, caused by the sun, while the second is not because the clouds don't derive from the house on the hill (it's just a random event) even if it looks like the shape of the house is drawn by the clouds:

Figure 9.1 – Isomorphic example (left) and non-isomorphic example (right)

A bit more formally, we can say that *homomorphic encryption* is a form of *encryption* that allows *computation* on *ciphertexts*, generating an encrypted result that, when decrypted, matches the result of the operations as if they had been performed on *plaintext*.

Let's look at the scheme of homomorphic encryption:

Encryption ⟶ [Operations on Encrypted Data] ⟶ Encrypted Result ⟶ Decryption

Figure 9.2 – Flow of operations in homomorphic encryption

This scheme can be executed by performing mathematical operations, such as multiplication or addition. If the system performs all the mathematical and Boolean operations, in this case, we say that the algorithm is *fully homomorphic*; otherwise, it is *partially homomorphic*. For instance, RSA or other algorithms, such as **ElGamal** or **MBXI**, are partially homomorphic. In RSA, we can find the homomorphism of multiplication, as we will discover in the next section.

Partial homomorphism in RSA

Let's take the RSA algorithm to explain this correspondence between ciphertext operations and plaintext operations (in this case, multiplication).

We know that RSA's encryption is as follows:

```
[M]^e ≡ c (mod N)
```

Here, we have the following:

- `[M]` is the message.
- `(e)` is the public parameter of encryption.
- `(N)` is the public key composed by `[p*q]`.
- `(c)` is the cryptogram.

Now, we are supposed to have two messages, `[M1]` and `[M2]`, and we encrypt them using the same public parameters `(e, N)` to generate two different cryptograms `(c1, c2)`. The result of the encryption will be as follows:

```
c1 ≡ M1^e (mod N)
c2 ≡ M2^e (mod N)
```

If we multiply the ciphers `(c1*c2)`, we get as a result a third cryptogram `(c3)`, such that the following applies:

```
c3 = c1 * c2
c3 ≡ M1^e (mod N) * M2^e (mod N)
c3 ≡ M1^e * M2^e (mod N)
```

I have simply substituted the operations on encrypted messages `[M1]` and `[M2]` with the cryptograms `(c1)` and `(c2)`.

All these operations can be regrouped into one cryptogram `(c3)`, such that we have the following:

```
c3 ≡ c1*c2 (Mod N)
```

Or, we may even have the following:

```
c3 ≡ (M1*M2)^e (mod N)
```

Hence, we obtained a result of (c3), which is the encryption of the multiplication performed on the two cryptograms (c1*c2) and, at the *second level*, the operations performed on the messages, [M1*M2]. Let's now see what is meant by the *second level*. Decrypting (c3) with the [d] private key, we obtain the same result as [M1*M2]:

```
c3^d (mod N) ≡ M1*M2
```

We can verify with a numeric example (because I suppose you could be skeptics) that the following applies:

```
c1*c2 ≡ (M1*M2)^e (mod N)
```

Or, we could also do the following:

```
c3^d (mod N) ≡ M1*M2
```

This means that if Alice decrypts a cryptogram (c3) with her private key, [d], she gets, with RSA, the result (in linear mathematics) of the multiplication of the messages, [M1*M2], previously performed by Bob and sent by him, hidden, to the unique cryptogram (c3).

This is called *partial homomorphism* of the multiplication in RSA.

Let's look at a numerical example to understand this better.

Let's take some small numbers as an example for verifying RSA partial homomorphism:

```
M1 = 11
M2 = 8
e = 7
p = 13
q = 17
N = 221 [p*q]
```

Compute the c1 and c2 cryptograms:

```
c1 ≡ 11^7 (mod 221) = 54
c2 ≡ 8^7 (mod 221) = 83
```

Now we calculate (c3) in such a way that we consider it to be our *first level*:

```
c3 ≡ c1*c2 (mod N)
```

We then calculate (c3) in another way that we consider to be the *second level*:

```
c3 ≡ [M1* M2]^e (mod N)
```

So, in terms of numbers, we have the following at the *first level*:

```
c3 ≡ 54 * 83 = 62 (mod 221)
```

We have the following at the *second level*:

```
c3 ≡ [11 * 8]^7 (mod 221) = 62
```

As you can see, we found a degree of correspondence between the operations (multiplications in this case) performed with cryptograms c1 and c2 and multiplications performed on messages [M1, M2].

Up to this point, it seems that everything is normal. We have only applied simple mathematical substitutions to the equations performed with both cryptograms and messages.

However, another important correspondence turns out to be the decryption of (c3), which is what makes the difference. In fact, Alice is able to decrypt the two cryptograms transmitted by Bob (c1, c2) through her unique private key, [d]. Using the well-known Reduce function of the Wolfram Mathematica software to find [d], the private key, we have the following:

```
Reduce [e*d= 1, x, Modulus -> [13- 1] [17 - 1]]
d = 55
```

But now, the surprising part is that by decrypting (c3), the result of the multiplication of (c1*c2), Alice can also ascertain the result of multiplication of the messages, [M1*M2].

Remember that c3 = 62 and [d] = 55, and we have that the decryption function in RSA as follows:

```
c^d ≡ M (mod N)
```

In this case, we have the following:

```
c3^d ≡ M1*M2 (mod N)
```

Substituting the numbers gives us the following:

```
62^55 ≡ 88 (mod 221)
```

In fact, it turns out that we have the following:

```
M1*M2 = 11* 8 = 88
```

This is the result we wanted to obtain!

As you can see, even the RSA algorithm holds some homomorphic properties, including multiplication. However, this is just a case study because we need to perform homomorphic searching differently from how RSA goes about it. So, let's analyze homomorphic encryption a little bit deeper and its implications for data security and privacy.

Analysis of homomorphic encryption and its implications

Analyzing the partial homomorphic scheme proposed for RSA in the preceding section, we can see an interesting correspondence between the operations on cryptograms (data expressed in clear) and messages (data in blind).

This correspondence is just what we call homomorphism.

Now, the problem is that RSA, just like most of the algorithms explored until now, is *partially homomorphic* and can only represent some mathematical operations, such as multiplication or addition. The real difficulty is finding an efficient algorithm that represents all the mathematical and Boolean operations together.

Another simple example (case study) of how a form of homomorphism can be represented is performed by addition.

Let's take [A] and [B], two secret values that give [C] as their sum:

```
A + B ≡ C (mod Z)
```

Now, let's take two encrypted value correspondents respective to A and B.

For example, the encrypted values are a = 9 and b = 2, the sum of which is c = 11.

We can generate a homomorphic partial (sum) correspondence such that we have the following:

```
A = 87 —> a ≡ 9 (mod 13)
B = 93 —> b ≡ 2 (mod 13)
C = A + B = 180 —> c = a + b ≡ 11 (mod 13)
```

We can also represent these operations in a row, so you probably now have a better idea of what homomorphic means:

$$C = 87 + 93 = 180 = 11 \text{ (mod 13)}$$

$$c = 9 + 2 = 11$$

Figure 9.3 – Correspondence of the homomorphic sum between plaintext values and encrypted values

As you can see, this simple correspondence makes it possible to operate in *blind* with the encrypted values, $a + b = 9 + 2$, and expose the result $c = 11$, but it can preserve the *privacy* of the numbers, $A = 87, B = 93$, and $C = 180$, exposing only an isomorphic value, $c = 11$, which is not the real value of C but represents its correspondent (isomorphic) value. The condition is that the function that transforms the value is a one-way function, so it will be hard to return from (a) to [A], (b) to [B], and the result (c) to its isomorphic correspondent [C].

I have deliberately highlighted the term *privacy* (I know it may appear strange to talk about the privacy of a number). It is just that the result we want to obtain while operating with such forms of isomorphism preserves data privacy. Operating with data in blind, indeed, we obtain a result that can be exposed clearly without the fear that any element of the equation can be discovered.

I started to conceive the project to implement a CSE, a search engine able to operate with data in blind, not *conscious* of what it is searching for, and agnostic of any algorithm used to perform the encryption/decryption of the data retrieved.

Before presenting CSE, I want to analyze a little bit of the math behind search engines (for general purposes and related to CSE) to understand better how search engines are implemented and how fascinating they are.

Figure 9.4 – Searching in blind

Now it's time to explore the math and logic behind traditional search engines and understand how complex it is to perform an efficient search on encrypted data.

Math and logic behind search engines

Everything related to the *network problem* and network theory started from Konigsberg's bridges dilemma in 1735. Euler was the first mathematician who solved a dilemma (relating to an apparently simple question) about crossing a set of bridges over the river Pregel, in the ancient city of Konigsberg in Prussia, known today as Kaliningrad.

As you can see in *Figure 9.5*, seven bridges link the river's two banks (A and B) through two islands (C and D).

The question is: is there any way to cross all seven bridges of Konigsberg by going over them just once?

You can try to find a solution by yourself before going on:

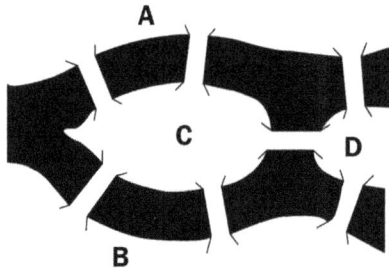

Figure 9.5 – The seven bridges of Konigsberg

The key is to recognize that what we are facing is a network (made of bridges), and we have to draw it to understand the solution. Compared to the seven bridges network, there are only four points where a wayfarer can be located at a specific moment, as we can see in the following diagram:

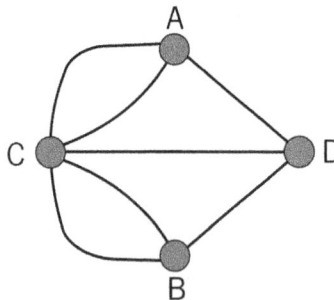

Figure 9.6 – The network of Konigsberg's bridges

Euler demonstrated that the question of crossing all seven bridges by going over them just once has no solution. That is because the network of bridges has four hubs (points), corresponding to seven links (bridges), and when the number of hubs is even while the number of links is odd, the problem has no solution.

The issue of crossing networks efficiently is one of the most important elements to consider if you want to design and implement an efficient **search engine** or **social network** or perform **big data analysis**. We have to find a way to retrieve the instance (the searched word, for example) and then cross the network and return with the least effort possible.

A similar problem is represented by searching for the best itinerary among many choices represented, for example, by different roads on a map. Have you ever wondered why Google is a champion not only when it comes to finding any answers to your questions for you but also for finding the best solution regarding an itinerary covering hundreds of miles and finding the destination in a few milliseconds?

The problem is related to *circuits*, in particular **Eulerian circuits** and other similar mathematical problems.

To go deeper into this question, let's see an example of how a network works, an example of which is represented in the next figure, called **Papillon** for its particular form:

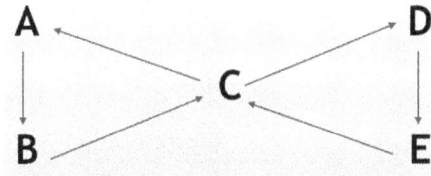

Figure 9.7 – Eulerian Papillon network

The Papillon network has a Eulerian circuit of this type:

A—>B—>C—>D—>E—>C—>A

In this representation, as you can see if you experiment by yourself, we pass by C twice (remember that we started at A, but it works the same if we start at C or any other point).

By changing the path a little bit, crossing the network through the same five hubs – A, B, C, D, and E – is possible without passing point C twice, as you see here in *Figure 9.8*:

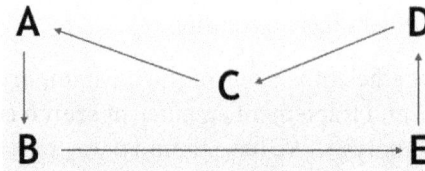

Figure 9.8 – Eulerian cycle

In this case, the circuit is **A—>B—>E—>D—>C—>A**.

These problems send us back to the following question: how do we find the best itinerary among a circuit made of hubs?

If we want to implement an efficient search engine and ask it to find the best itinerary to cross over a group of hubs (as we have previously represented in *Figures 9.7* and *9.8*), we should impose some conditions. As you know, in real life, sometimes the constraints are expressed by traffic or by avoiding highways or tolls, but most commonly, the first condition is represented by the shortest itinerary of the path. In our case, the first condition is to find the shortest path to complete the cycle among all the points, pass by each point only once, and move in a clockwise direction as the last condition. Our search engine has to calculate the shortest distance starting (for example) from A and returning to A, intersecting all five points just once and moving clockwise. In this case, the representation explained in *Figure 9.9a* is a solution (but not the only one).

Another solution is A—>B—>C—>E—>D—>A, proposed in *Figure 9.9b*. The scheme proposed in *Figure 9.7* (Papillon) doesn't respect the second *rule* we gave to our search engine, so it is not valid.

To find the best solution, we have to assign a value to each segment linked by the lines that represent the itinerary that our search engine will run. Suppose that a segment (for example, A—>B) has the following values:

```
A->B = 1
B->E = 3
A->C = B->C = C->E = D->C = 2
```

The itinerary's values shown in *Figures 9.9a* and *9.9b* are equally efficient.

Itinerary (IT$_1$) has a value of $1 + 3 + 1 + 2 + 2 = 9$.

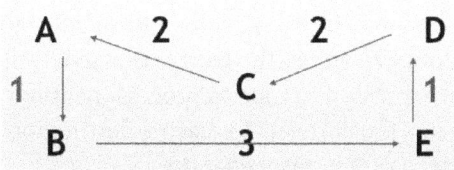

Figure 9.9a – IT$_1$ – a possible solution for the cycle

Our itinerary is equally efficient as itinerary (IT$_2$), as you can verify in *Figure 9.9b*:

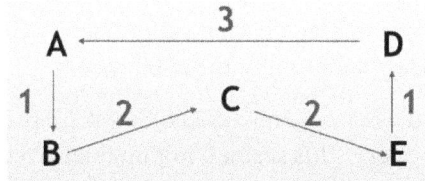

Figure 9.9b – IT$_2$ – another possible solution

Itinerary (IT$_2$) has the same value as IT$_1$: $1 + 2 + 2 + 1 + 3 = 9$.

So, the two solutions, IT$_1$: A—>B—>E—>D—>C—>A and IT$_2$: A—>B—>C—>E—>D—>A, have the same total values, IT = 9, which is the most efficient for our search engine.

Now, suppose we want to calculate the total value expressed in the Papillon circuit (*Figure 9.7*). In that case, you will discover that it is less efficient to run the entire circle (the total value of Papillon is IT$_p$ = 10), but it's the best scheme *to search for* the point (C) and return to A in less time and length, as you can see in *Figure 9.10*. In fact, in the Papillon circuit, we can return directly to A from C: IT$_3$: A—>B—>C—>A = 5.

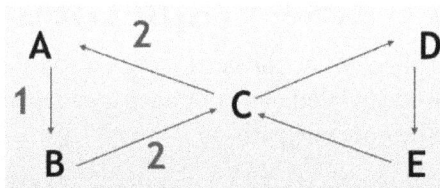

Figure 9.10 – The itinerary IT$_3$

In these graphs, you start to understand the logic behind a search engine. One of the principal requirements of a search engine is to be efficient. We have to keep in mind that whenever we project a search engine, be it encrypted or not, efficiency is an important element. Just like traveling along the street, the path expressed in miles or another metric always has a correspondence expressed in time to process the itinerary. That is the same as when you ask Google to process the best way to reach a destination and choose as a variable the travel time instead of the travel distance.

Time is the primary variable to consider when we launch a query on a search engine. It's agreed that 2 seconds is the maximum time limit to answer a query (in an unencrypted way). In other words, for example, a user is not prepared to wait more than 2 seconds for an answer while searching for something on their cellphone. It's for that reason that we have to project a circuit with the maximum grade of efficiency.

One of the constraints adopted when we embarked on the CSE project was to stay below 2 seconds of elapsed time per query. This seemed impossible, given similar (but not identical) previous encrypted searched engine projects such as TOR: `https://www.torproject.org/`.

When we deal with encrypted data, the latency of the communication becomes heavier. This is because mathematical operations (as we have seen in this book) have more computational power than operations performed on plaintext. Multilayers of encryption (as in TOR) do not always reinforce the system but certainly make it computationally heavier.

Always remember that in cryptography, the weakest link in the chain destroys the entire chain. In the next section, we will examine another essential element in graph theory that is crucial for the implementation of a search engine: tree graphs.

Introduction to trees – graph theory

In graph theory, a tree is a graph in which the vertices (v), also called nodes or points, are, in some sense, related. Each of the related pairs of nodes is called an edge (or link or line) and two nodes are connected by only one path.

In the following figure, you can see a tree where (1) is the root of the tree, 2, 3, … 7 are the nodes, and 8, 9, … 15 are the leaves. The segments that link each of the points are the edges.

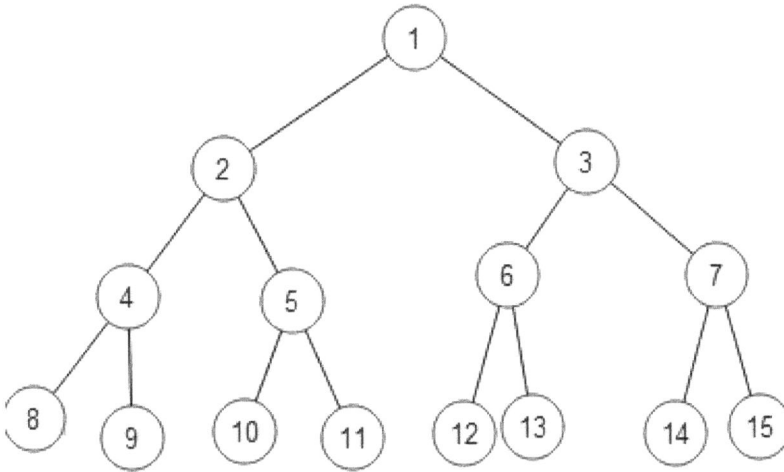

Figure 9.11 – A tree graph

We have seen many cryptographic algorithms in this book. The main scope of these algorithms is related to data security. However, the main scope of the algorithms we are now approaching, also called *codes*, not only concerns security but also aims to preserve the information in a synthetic and usable way. This concept is related to retrieving information in the most efficient way possible. Like the problem of the bridges of Konigsberg and Eulerian circuits, tree graphs can be used to calculate the best way to retrieve information.

This problem also goes by the name of *prefix codes* related to *lossless data compression*. These two elements are both very important in cryptography concerning codes and data compression. In this particular case, we will see Huffman coding in the next section, a kind of lossless data compression in which it's possible to retrieve the information produced without losing any data.

The prefix codes that we will analyze now are important because they allow a word to be encoded uniquely, as in the following example:

character	a	b	c	d	e	f
code	0	101	100	111	1101	1100

Table 9.1 – An example of prefix codes on letters

The encoding of the abc string, for example, will be 0 · 101 · 100. We are sure in this way that if we are searching for a string starting with abc, we don't find any code other than 0101100. This method is similar to a telephonic prefix index; we assume that the index we are calling is unique based on a country, region, or city. So, when we call a prefix number starting with 001, for example, we are dealing with North America. By adding the next prefix, 415, to 001, we are calling a number in California, specifically, the San Francisco area. In this way, telephone companies can divide the globe into prefix code numbers: 001 for America, 002 for Africa, and 003 for Europe, as you can see in the map in *Figure 9.12*. The result is an easier way to search for a number located in a country-region-city code.

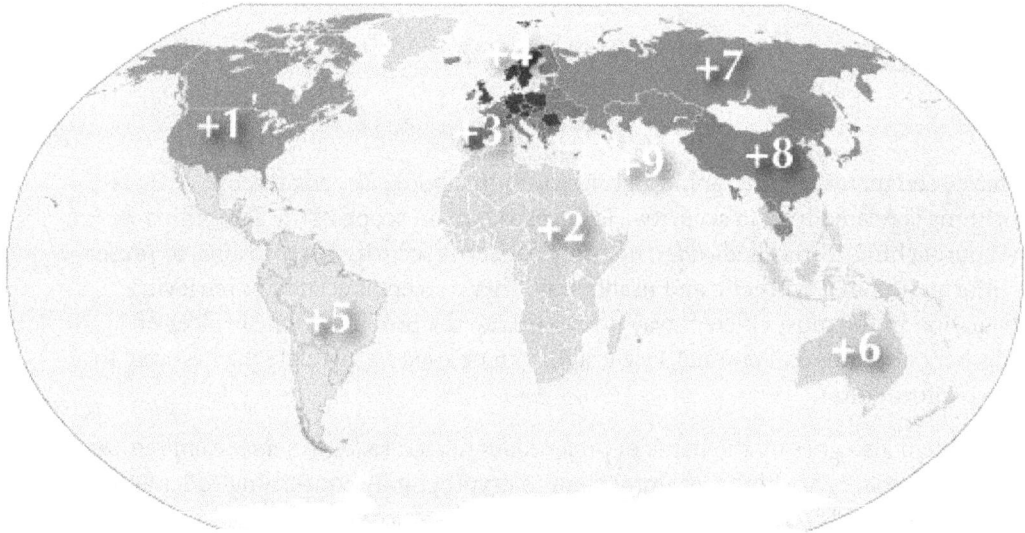

Figure 9.12 – Prefix number codes across the globe

A search engine works similarly. Just like prefix numbers, the indexing algorithms have to be very efficient and accurate to search inside billions of instances, retrieving the information requested in a few milliseconds. Even if this book's scope is cryptographic algorithms, I want to go a little bit deeper to explain how tree graphs work because they are key to implementing an efficient search engine for encrypted data. So, let's go and implement a tree graph.

Huffman code

Suppose we want to solve the following problem: finding a way to represent text in binary code in the shortest way possible. It will be optimized, reducing the number of strings to as few as possible to encode text.

We have already seen a way to encode letters, numbers, and notations with ASCII code in *Chapter 1*, *Deep Diving into Cryptography*, in the *Binary numbers, ASCII code, and notations* section. But here, the problem is different: we want to be much quicker and more efficient in terms of encoding the text using as few bits as possible.

So, we can use an elegant method ideated by David Huffman, which operates by implementing a special tree graph.

There are 26 letters in the English alphabet, but not all the letters hold the same frequency in a text. We want to implement a tree graph that is able to codify, for example, six letters – a, o, q, u, y, and z – whose relative frequencies in a given text are as follows:

```
a = 20
o = 28
q = 4
u = 17
y = 12
z = 7
```

The objective is to discover the minimum possible number of bits, (0) and (1), to codify text made up of these letters.

Take a look at *Figure 9.13*, where you can find the entire strategy of this tree graph implementation:

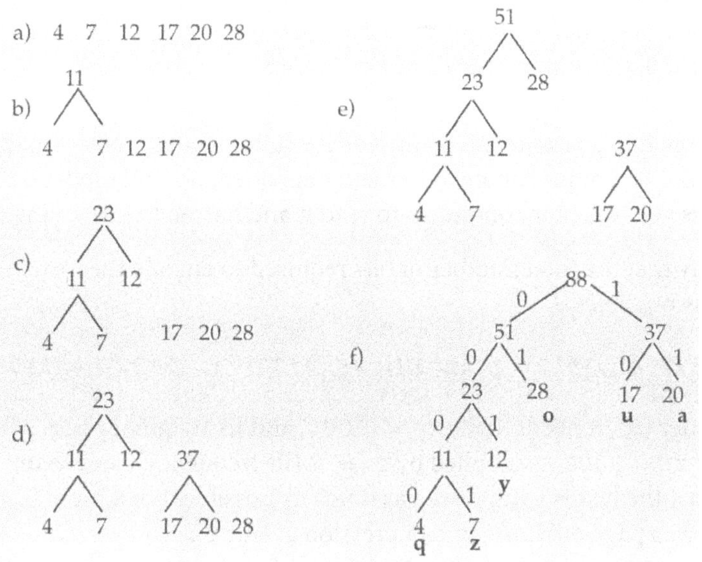

Figure 9.13 – Huffman tree graph implementation

Now we are going to analyze the sequence of steps necessary to implement the graph:

- As the first thing, **a)**, we form a sequence of nodes, six in this case, disposing of them in ascending order based on the frequencies of the letters.

- In stages **b)**, **c)**, **d)**, and **e)**, you can see the implementation of the main tree, where the nodes are partial trees of the same big tree, **f)**. They are composed of the sum of the number pairs, starting from the smallest, linked pair by pair until it reaches the top of the tree (see the final step, **f)**). Finally, we find the summit of the tree represented by the number **88** (root), which is the sum of the vertices **51** (composed of 23 + 28) and **37** (17 + 20). Now that we have completed the tree, we can assign the value **0** to all the left edges and **1** to the right edges (it also works in the opposite mode). Moreover, we assign to each letter a corresponding number related to its frequency: **q** = 4; **z** = 7; **y** = 12; **o** = 28; **u** = 17; **a** = 20.

- After the tree has been built, we can read the *string code* for each letter (that is, for the letter **z**) starting from the root and add a 0 every time we go left to a child, and a 1 every time we go to the right. So, for example, to read the letter **z**, we start from the summit root **88** and count down how many instances of 0 are on the left and how many instances of 1 are on the right-hand edges we encounter before reaching the letter **z**, which is 0001, in the following way:

```
q = 0000, z = 0001, y = 001, o = 01, u = 10, a = 11
```

> **Important Note**
>
> It is possible to use other schemes to encode the tree. For example, we can assume that 0 will be on the right and 1 on the left.

The structure of the tree guarantees that any code will not be the prefix of another one. For example, a = 11 will be unique for any other letter encoded; no other letter will start with 11. Therefore, it is very easy for computers to read it and hasten the decoding process.

We can also verify that the total number of bits required to encode the text, in this case, will be as small as possible:

```
(4*4) + (4*7) + (3*12) + (2*17) + (2*20) + (2*28) = 210
```

To encode the letter (q), it needs 4 bits, q = 0000, and its frequency is q = 4; the same bits for z = 0001, multiplied by z = 7 (its frequency), and so on. We will make sure that the text is compressed as much as possible. For a more in-depth insight, visit the web page of Huffman compression at https://www2.cs.sfu.ca/CourseCentral/365/li/squeeze/Huffman.html.

This prefix and compression code can help us significantly to compress the text and search on it, but these are not the only elements we must consider when implementing an efficient encrypted search engine.

Hash and Boolean logic

Another key element that needs to be to be considered is the *hash functions* that have to be combined with tree graphs. We have already covered hash functions in *Chapter 4, Introducing Hash Functions and Digital Signatures*, of this book. In the same chapter, we have also seen more Boolean operators useful for hash function in the section: *Logic and notations to implement hash functions*. However, while in cryptography hashes are employed for a digital signature to avoid exposing the content of the message, [M], here we use hashes for a different scope. The *object of a query* is also called the **keyword**. Another definition of a keyword could be the result of a process of indexing a string or a group of strings in a database.

If our search engine is efficient, it will be indexing (a similar process to encoding) as few strings as possible, recognizing and discarding the double-indexed ones. In a search engine that works with encrypted data, the query itself is encrypted too.

In CSE, it's possible to search for complex queries made by multiple keywords. So, it's necessary to use Boolean logic operators such as OR, AND, and NOT, which we have already seen in *Chapter 2, Introduction to Symmetric Encryption*, in the section entitled *Notation and operations in Boolean logic*.

For example, if we want our search engine to retrieve the part of this book that contains this section, we will perform the following query:

```
[Hash and Boolean logic]
```

Once the whole content of the book has been encrypted and divided into chapters (we can suppose that each chapter is a file that is encrypted), we expect to receive an answer with the file (or files) that contains the three keywords (Hash, Boolean, Logic), linked in this case by the AND Boolean operator:

```
[Hash "AND" Boolean "AND" logic]
```

In other words, it means to retrieve all the files (chapters) that currently contain the Hash + Boolean + logic keywords together (excluding, for example, the chapters containing either of the words Hash, Boolean, or logic).

In this case, the proposition made by multiple keywords linked together by the AND Boolean operator will return chapter 9, specifically this section, in an encrypted way. Now, with our private decryption key, we can decrypt and read the section's content.

That is how CSE works. Moreover, in our search engine, it's possible to set up search operations among encrypted data using all three operators, AND, OR, and NOT.

So, we can search, for example, in encrypted content, for a query such as the following:

```
[Boolean "AND "Logic "NOT" Hash]
```

This means that the search engine will search for all instances containing the words Boolean + Logic together, excluding Hash. In this case, we will be referring to *Chapter 2, Introduction to Symmetric Encryption*, where we find the *Boolean logic* section.

Queries in CSE are case-insensitive; for example, if we type hash or Hash, the query results will be the same. Finally, you should have noticed that I have [bracketed the query] because even queries themselves on CSE are encrypted for privacy and security reasons.

The secret to searching for a keyword efficiently is to combine the index as a tree with hash functions and Boolean logic to obtain the most efficient path to the *target keyword*.

Our research team at Cryptolab has invented a unique technique for finding an *encrypted keyword* in the content of unstructured encrypted files in an average of 0.35 seconds semi-independently (-%) according to the number of bits processed.

So, it's now time to show how CSE has been created and what its principal functions are.

CSE explained

Data is often communicated across networks and stored on remote servers, which may be unsecured and non-private; eventually, it needs to be browsed, searched for, and manipulated regardless of the location where it is held. In the case of sensitive data (for example, in healthcare), it needs to be kept secure and private throughout the process. State-of-the-art technology regarding sensitive data management on remote servers achieves this objective via sub-optimal combinations: sensitive data is usually made secure by local encryption and then communicated and remotely stored. In the event of requests to browse or search, it is decrypted on the remote server and then accessed. If manipulation is requested, additional encryption may even be necessary. This combination is functional but sub-optimal as it wastes computational power and it exposes sensitive data in a clear-to-read form on remote servers (which are often provided by third-party cloud services). This problem could be prevented by employing the back transmission of encrypted data to local storage, where data would be safely decrypted, browsed, searched for, eventually encrypted again, and forward-transmitted to remote servers. It is clear how this solution protects sensitive data, but it results in longer processes, wastes bandwidth, and exposes data to multiple communications over unsecured networks.

In practical terms, this results in two very common behaviors: on the one hand, some private users renounce their privacy to take advantage of services and cost savings (this is typically the case with cloud services related to smartphones, which are appealing and inexpensive as long as the user is willing to share their private data). On the other hand, some institutions need to manage sensitive data, and since they cannot compromise in terms of privacy, they maintain privately managed data centers. When resilience to disasters is a requirement (typical for critical infrastructures in general), such data centers must additionally offer geographical and technological redundancy, thereby increasing costs even further.

This introduces the idea of *encrypting the data before outsourcing* to the servers and *always keeping the data encrypted*; thus, a potential hacker who overcomes all the security barriers and steals data cannot read it. The technical problem with encrypted data is that there are black boxes, and it's not possible to apply operations on encrypted data without decrypting it first. For this reason, it's a common practice to put this kind of data in a secure environment (such as a private data center) where it's possible to decrypt without exposing it to hackers who can steal the data. This scenario is also not completely secure because a *trusted* person can access data, read it clearly, and copy and paste it to another media, such as a USB key.

In the following figure 9.14, you can see a scheme of encrypted data in external storage. As you may notice, the encryption keys (the data key and the master key) are encrypted and stored in the external server (for example, the cloud) together with the encrypted data storage. The keys are used to unencrypt the data in encrypted storage when a query is requested. This process is necessary during normal encryption to perform searching in the database. The shortcoming with this system is demonstrated when the data is decrypted to allow searching, thereby exposing the content.

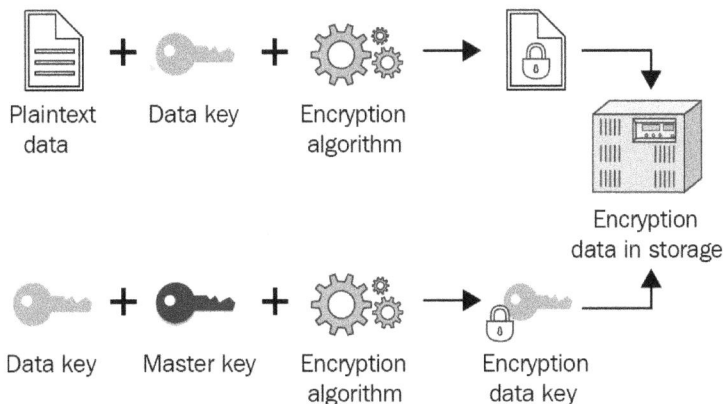

Figure 9.14 – Encrypted data storage in the traditional way (not efficient and not secure)

Replacing this method with a method of searching in blind makes the system wholly efficient and secure. That is what CSE can do, so let's see how it works and what its features are.

The innovation in CSE

CSE represents a technological solution based on isomorphism, *a transformation that preserves information*, as it can offer searching, browsing, and manipulation on encrypted data, that is, with *zero knowledge*; encrypted, sensitive data stored on public cloud providers will eventually be fully accessible by data owners only, still allowing cloud providers to offer searching, browsing, and manipulation over encrypted data by means of *encrypted queries*, thereby preserving the zero knowledge. CSE is able to operate independently according to the kind of encryption that the user decides to adopt. On the one hand, private users could take back legitimate ownership of the value of their data and eventually trade it back to service providers through a fair deal, which is very different from the *take-it-or-leave-it* approach that we currently witness; this is a key value for the next generation of digital citizens who could also exchange or sell their data autonomously. On the other hand, public administrations, institutions, and critical infrastructures would be permitted to use a multiplicity of public cloud providers for safe and redundant data storage, with an important impact in terms of reduced costs and enabling access to tools for big data analysis.

The searching, browsing, and manipulation of encrypted data is a unique and unprecedented feature that will disrupt any business model where sensitive digital data is generated, communicated, stored, and processed.

CSE might constitute the technological architecture for *next-generation* search engines; data would be publicly available and searchable while preserving full encryption and integrity. New search engines would be perfectly aligned to EU/US legislation on data privacy (GDPR) and recent socio-economic trends concerning data integrity, privacy, and protection. CSE scale-up will be vital for further deployment of the enabling infrastructure.

But let's now see how CSE works. In the scheme of *Figure 9.15*, you can find a good representation of the cycle accomplished by a query performed in an encrypted way that reaches encrypted content hosted, for example, by a public cloud provider:

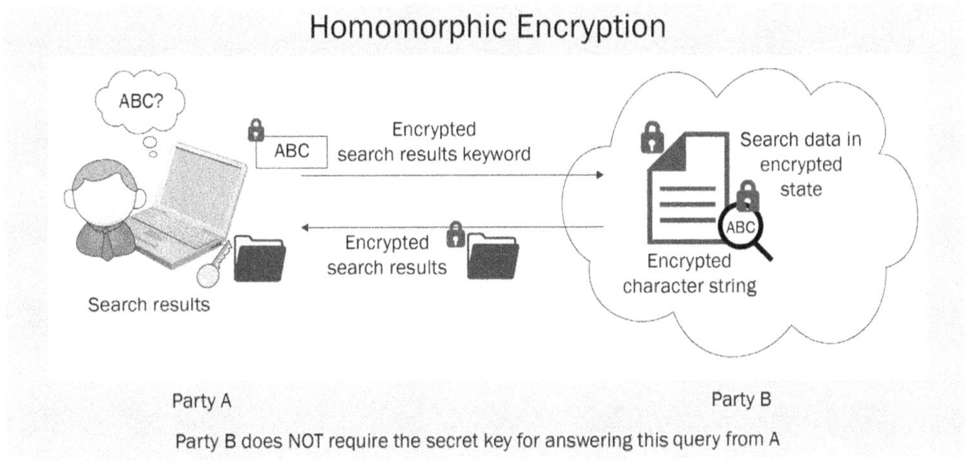

Homomorphic Encryption

Figure 9.15 – CSE scheme

In the scheme of *Figure 9.15*, there are two parties, but in reality, there is no exchange of information between the two subjects. Party A (here represented by a human) is generally a private server that hosts the encrypted keys and performs the query. The server asks party B (the cloud) to search in blind for a query inside the encrypted content.

In *Figure 9.16*, you can see a simple scheme of a hybrid cloud architecture applicable to CSE:

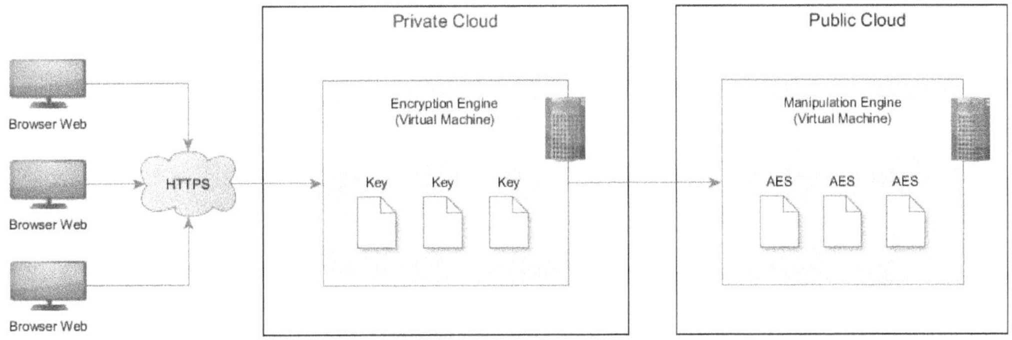

Figure 9.16 – Basic hybrid architecture of CSE

In this basic scheme, the two virtual machines, **Encryption Engine [EE]** (which is the private server) and **Manipulation Engine (ME)** (which is located in the public cloud), are connected to a symmetrical encryption algorithm. In CSE, the system is agnostic to the algorithm used to encrypt the files. In the first version of CSE, AES-256 has been adopted (explained in *Chapter 2, Introduction to Symmetric Encryption*), but in the next version, a quantum cryptography algorithm will be adopted (besides the classical cryptography) to send files through a quantum channel between the [EE] and (ME).

> **Important Note**
>
> I've used the brackets to indicate the type of environment used in the virtual machines:
>
> [EE] stands for private protected environment (i.e. a private cloud).
>
> (ME) stands for public cloud or a non-protected environment.
>
> Both the VMs have enrypted data stored inside them.

Generally, HTTPS (as you can see in *Figure 9.16*) is used only inside the protected private space to connect the user's devices to the [EE] located in a private cloud (that is, the company server, for example).

The [EE]'s primary function is to generate random encryption keys, all different, one by one, for each file encrypted. Hence, it will be impossible (-%) to find two files with the same encryption even if their plaintexts are identical.

The (ME)'s duty is to store the encrypted files and search through the encrypted content. In an efficient architecture, a search engine can be hosted in a third virtual machine entirely separated by the [EE] and (ME).

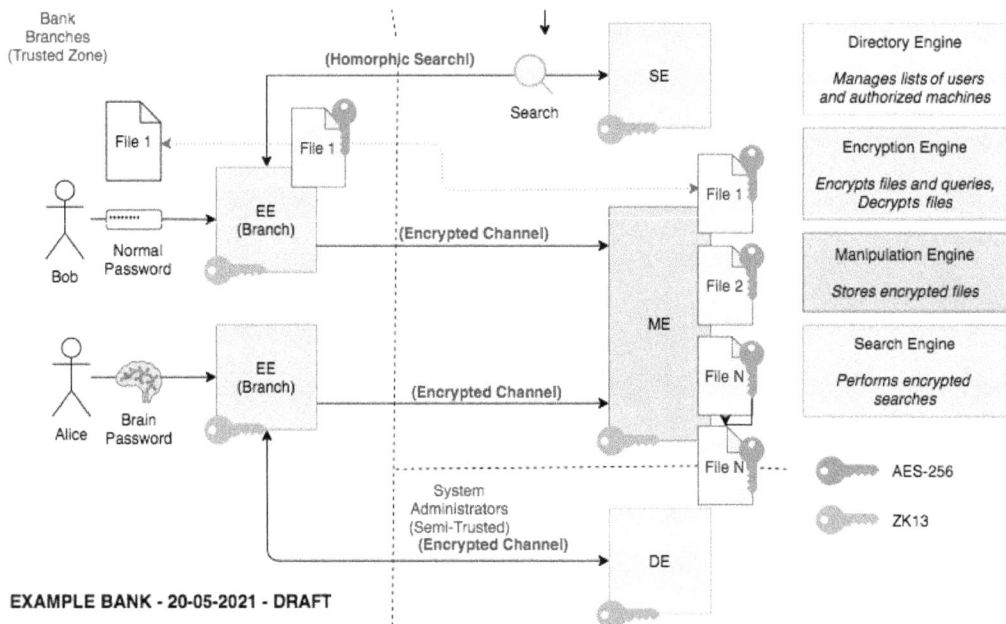

Figure 9.17 – A complex architecture scheme of multiples [EE] linked to (SE), (ME), and [DE] in CSE

In the preceding *Figure 9.17*, the CSE architecture comprises several [EE]s connected to several (ME)s identified through a zero-knowledge protocol, which can provide a *proof of identity* to the virtual machines' network. In the first version of CSE, a ZK13 non-interactive protocol was used (analyzed in *Chapter 5, Introduction to Zero-Knowledge Protocols*), which controls the identity of the virtual machines, thereby avoiding possible man-in-the-middle attacks. In a future version, a special quantum protocol of authentication will be used to identify the virtual machine in CSE. In *Figure 9.18*, you can see a screenshot of my private *encrypted box* web interface on CSE:

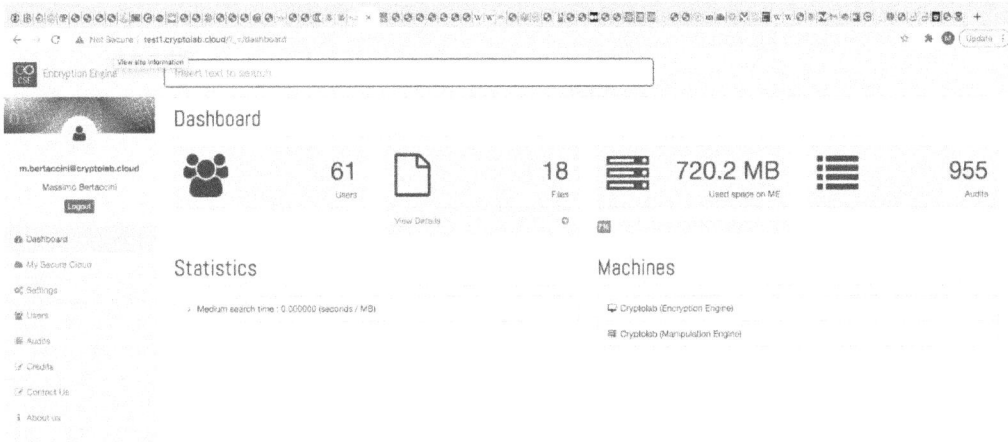

Figure 9.18 – Web interface of CSE

Now that we have seen a working schematic of CSE, we will discover its computational grade of efficiency next.

Computational analysis on CSE

The secret to searching in blind is to combine elements such as hash functions, encryption algorithms, authentication algorithms, and user interfaces into one helpful platform.

Computationally, the problem is to combine the best grade of security, given by the algorithms of encryption, decryption, and authentication, as well as the policies to log in and manage the platform, with the efficiency of searching in encrypted content. Suppose you want to search blindly inside the entire content of Wikipedia, encrypted, for example. In that case, the problem is how big the amount of content (expressed in kilobytes, gigabytes, or terabytes) is and the number of *keywords* processed. In other words, we have to face off with a complex system given by the sum of the elements combined into CSE:

- Number of files encrypted
- Time to encrypt the files
- Bits per file and bits per total content processed
- Number of keywords processed
- Elapsed time per query
- Human-to-virtual machine authentication
- Virtual machine-to-virtual machine authentication
- Time to decrypt the file searched
- Time to wipe out the content from the memory of the system

All these elements have to be combined in an optimal solution given by an equation that I have processed to evaluate the **efficiency (E)** of a CSE:

E = S/T

Here, we have the following:

- E: Efficiency of the system
- S: Security of the cryptographic algorithms
- T: Time processed

The equation says that the higher the system's security level, and the less time taken to process a query, the more efficient the system will be.

In the equation proposed, (0) is the minimum security level of an algorithm and 1 indicates 100% security. Let's suppose now that the system doesn't adopt any encryption to protect the data, so then we have **security (S)** = 0, as opposed to if we use quantum cryptography encryption, where we would probably have a very high security level, S = 1. The searching time, T, is a variable that goes from (0+) to infinity, even if we know that an acceptable searching time answer is at most 2 seconds per query. Hence, the maximum degree of efficiency will be very high security, for example, S = 0.999999, and the lowest time per query as possible, T = 0.00001. So, if we want to attain the highest efficiency, we should augment security and reduce time.

Note

Here, we suppose that data is transmitted and processed through the internet, so it can be exposed to external attacks and be spied upon. However, it's not said that a system is more protected because data is encrypted. If a system works offline and information is not shared or processed externally from the offline computer, the system could have a higher degree of security even if its content is not encrypted.

The (0+) notation stands for the lower limit of T that tends to zero, but cannot be zero.

Since time is a deterministic and estimated variable, the problem is how to estimate the best grade of security and degree of efficiency.

In CSE, AES-256 is adopted among the other algorithms as a symmetrical encryption algorithm to transmit the files between the [EE] and (ME). AES, as we saw in *Chapter 2, Introduction to Symmetric Algorithms*, is now a standard in symmetric encryption (valid at this time, although the same cannot be said for the future).

I have tried to calculate the grade of complexity of AES-256 bits in terms of attack and computation, related to the power of calculation at our disposal. Obviously, things are evolving fast, and what is considered secure now may not be in 3 or 5 years.

Let's try to follow the logical reasoning in the next section about how to calculate the efficiency of a cryptographic algorithm such as AES-256.

Example of computational brute-force cracking

Let's suppose we have a four-core machine, let's say a MacBook Pro 2015 i7 core. This machine equates to 1,024 MiB per second or 2^30 bytes per second.

As AES-256 uses a 16-byte block size (2^4), on average, a single high-performance PC can perform a calculation of 2^(30-4) = 2^26 blocks per second. Supposing that only one computer is always running for 60 sec * 60 min * 24 h * 365.25 days = 31,557,600 (seconds in 1 year), the total computation per year related to a single high-performance computer will be 31,557,600 * 2^26 = 2,117,794,686,566,400.

That is the number of keys we have to explore, approximately 2,117.8 trillion.

To perform a brute-force attack on AES-256, on average, you will need to try 2^255 keys.

> **Important Note**
> You could be lucky and try fewer than 2^255 keys, but it remains an incommensurable number.

So, we have to divide 2^255 by 2,117.8 trillion. The result is a number expressed as an exponent of 10, which is about 2.73 * 10^61. Written in full, this is 27,337,893,038,406,611, 194,430,009,974,922,940,323,611,067,429,756,962,487,493,203 years.

This number can also be represented as 27 trillion trillion trillion trillion trillion years. Just to give you a comparison, the Universe has only existed for 15 billion years!

What would happen if all the computers of the Earth were to join together to crack AES-256?

If you are passionate about computation or other strange questions about big numbers, you can try to use the *Wolfram Alpha* search engine and ask *how many computers exist on Earth?*

Alpha is a high-performance search engine supported by Wolfram Mathematica as a computational platform. You can check, and the answer you will receive is 2 billion PCs on Earth.

You can also check the entire process of my previous argumentation using Wolfram Alpha. Moving on, we divide the time employed by 1 computer by 2 billion, assuming that all the computers on Earth have the same computation of a four-core MacBook Pro i7. Obviously not, but I am optimistic.

The result will be 13,650,000,000,000,000,000,000,000,000,000,000,000,000,000,000, 000 years.

Or, expressed in a power of 10, it will be 1.365*10^52 years. This is an incommensurable computational time, even for all the computers on Earth.

Based on this thesis, we should assign the efficiency equation S = 1 to AES-256. But things are not that simple, and I will demonstrate why with a simple question:

What if, instead of combining all the power of classical computers, we use a quantum computer?

As we saw in *Chapter 8, Quantum Cryptography*, Shor's algorithm can crack the factorization problem. However, AES doesn't rely on this mathematical problem. So, isn't a quantum computer theoretically able to crack AES?

As you can probably imagine, cracking AES is a problem related to searching for something in a vast amount of content: the amount of 2^{255} keys hidden inside the algorithm. Grover's algorithm could be a good candidate (if it can be implemented into a quantum computer in the near future). Indeed, all algorithms, such as AES, that rely on hidden keys, basing their strength on the confusion and diffusion principle of Shannon, can theoretically be broken by a quantum computer.

> **Important Note**
>
> Grover's algorithm shows how to search in unstructured content with a quantum computer.

The good thing is that CSE is a system projected to be agnostic to any algorithm used. Theoretically, it can survive quantum attacks, replacing some algorithms inside its core system, even the *homomorphic search algorithm invented by Cryptolab's Team, HK16*, to perform the queries on the encrypted content. That is the core of the system will be replaced. It can be substituted by *quantum homomorphic searching*, which can work on a quantum computer more efficiently.

Finally, I wish to answer the question: what is the current efficiency of CSE?

It is not up to me to give CSE a score because I am one of the system's designers. But I can certainly say that given the algorithms adopted and the time elapsed per query with an average time of 0.35 seconds, CSE has superior efficiency both in terms of current security and time. That will remain valid until the conditions change, as we have seen.

Now that we have analyzed the efficiency of a system such as CSE, we are going to see the main applications of such a system in real life.

Applications of CSE

Most of the CSE address applications fall among **encrypted data storage** and **File Synchronization and Sharing** (**FSS**) in the cloud. As I have said, encrypted data storage in the cloud holds an important added value compared with other cloud applications: *homomorphic search*. An interesting use case of CSE applied to blockchain is *blind payments*. Here, the holomorphic search engine has the important function of listing the system's client anonymously and securely processing the transactions. In the healthcare field, two applications adopting CSE have already been tested: the management of **Electronic Health Records** (**EHRs**) in the cloud and **CovidFree**. Going a little bit deeper, CSE technology enables data search and big data analysis in the EHR system while preserving patients' security and privacy. CovidFree is an application patented in 2020 to fight the Covid-19 pandemic that maintains the privacy and security of data to monitor the virus. This application monitors and virtually contains the virus through the control of access in public places (airports, trains, buses, restaurants, and so on) without using any sensitive data to track people.

A scheme for sharing data in an encrypted environment is represented by the following diagram:

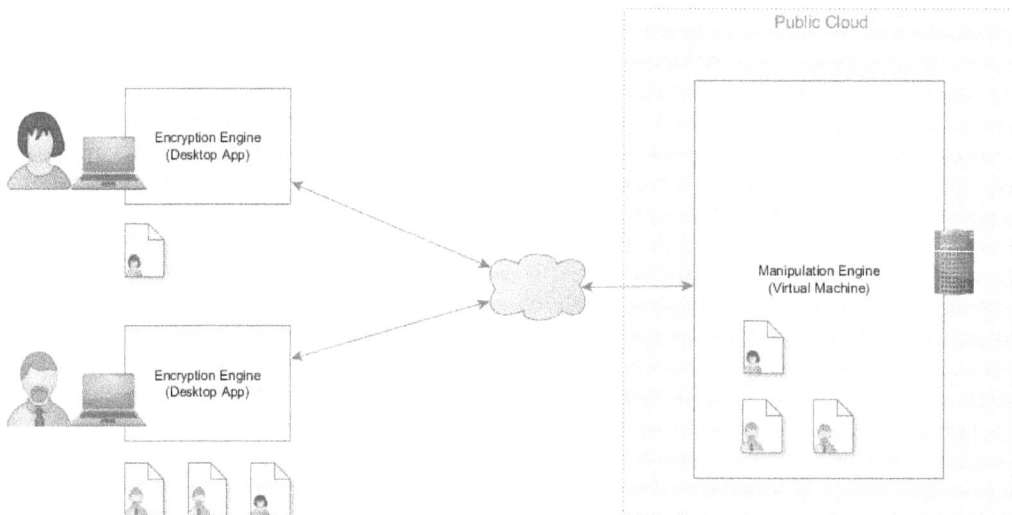

Figure 9.19 – Sharing data in the encrypted cloud

Other applications of CSE are related to video surveillance and video image recognition through AI. CSE technology, in this case, enables video management and real-time analysis of the encrypted data directly, thereby preserving the privacy of sensitive data within the video domain. As you can see in the following diagram of *Figure 9.20*, CSE applies to different fields in IT:

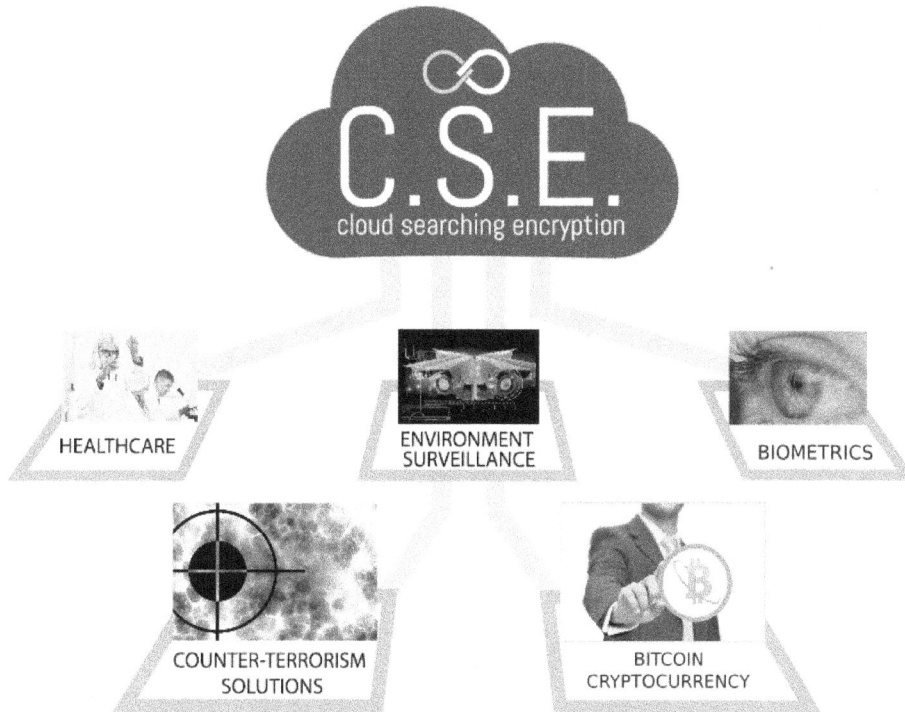

Figure 9.20 – Some fields of application of CSE

Another interesting use case is in the automotive field. In the future, more and more connected cars and autonomous cars will transmit data among themselves to understand, for example, when running on a particular route, whether an accident or a traffic jam will occur. In this scenario, it is easy for hackers to take control of the car if the data transmitted is in plaintext. CSE prevents this (or at least reduces the chances of it) because data is always transmitted and processed in an encrypted way. Even if hackers are sniffing around the communication channel, they won't find any readable messages, just strings of encrypted messages.

In biosciences, a scenario where CSE can be very useful is DNA storage and sequencing. This is a very sensitive field that combines the possibility of discovering sickness and many characteristics of a person, so it has to be managed very carefully. Like all data related to our wellness, physical, and psychological information, CSE can store the data collected in the DNA of a person and search for it in blind, preventing it being stolen or the information being spied upon by an unauthorized party.

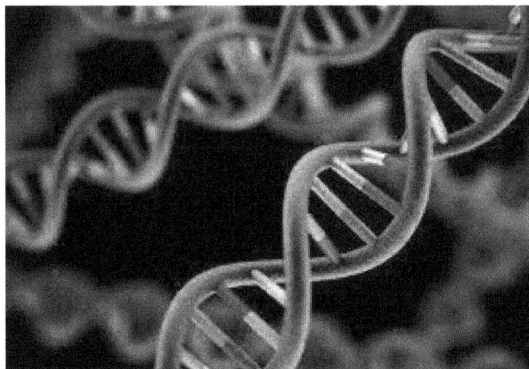

Figure 9.21 – DNA storage can be processed in blind using CSE

Most of the applications described here have already been tested and some are in use in real life. The hope is that the next generation of search engines will guarantee privacy and security to everyone, and CSE could be an example.

Summary

In this chapter, we introduced the basic principles of homomorphic encryption and its fundamental bases, and we analyzed the partial homomorphism in RSA.

After that, we explored the math behind search engines and, in particular, Eulerian circuits. We then undertook a deep dive into trees (graph theory), experimenting with a particular tree graph construction such as Huffman code. If usefully combined with hash trees and other elements, this algorithm is an excellent candidate for searching and compressing data in unstructured content.

Finally, we analyzed CSE, looking at its principal elements and possible future applications.

You have now learned what CSE is and how it is implemented.

Now that you have learned the CSE fundamentals and have explored most of the main cryptographic algorithms, I wish to congratulate you. I hope that this book will be helpful in your career and your professional life.

Index

Packt>

Packt.com

Subscribe to our online digital library for full access to over 7,000 books and videos, as well as industry leading tools to help you plan your personal development and advance your career. For more information, please visit our website.

Why subscribe?

- Spend less time learning and more time coding with practical eBooks and Videos from over 4,000 industry professionals

- Improve your learning with Skill Plans built especially for you

- Get a free eBook or video every month

- Fully searchable for easy access to vital information

- Copy and paste, print, and bookmark content

Did you know that Packt offers eBook versions of every book published, with PDF and ePub files available? You can upgrade to the eBook version at packt.com and as a print book customer, you are entitled to a discount on the eBook copy. Get in touch with us at customercare@packtpub.com for more details.

At www.packt.com, you can also read a collection of free technical articles, sign up for a range of free newsletters, and receive exclusive discounts and offers on Packt books and eBooks.

Other Books You May Enjoy

If you enjoyed this book, you may be interested in these other books by Packt:

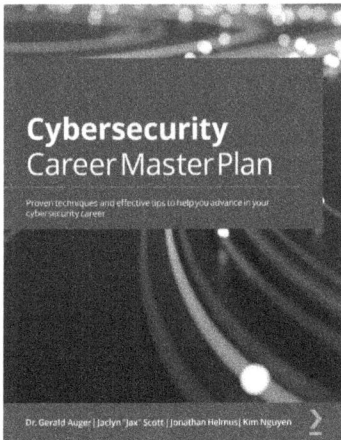

Cybersecurity Career Master Plan

Dr. Gerald Auger | Jaclyn "Jax" Scott | Jonathan Helmus | Kim Nguyen

ISBN: 978-1-80107-356-1

- Gain an understanding of cybersecurity essentials, including the different frameworks and laws, and specialties
- Find out how to land your first job in the cybersecurity industry
- Understand the difference between college education and certificate courses
- Build goals and timelines to encourage a work/life balance while delivering value in your job
- Understand the different types of cybersecurity jobs available and what it means to be entry-level
- Build affordable, practical labs to develop your technical skills
- Discover how to set goals and maintain momentum after landing your first cybersecurity job

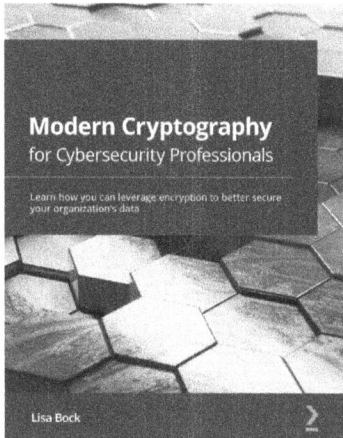

Modern Cryptography for Cybersecurity Professionals

Lisa Bock

ISBN: 978-1-83864-435-2

- Understand how network attacks can compromise data
- Review practical uses of cryptography over time
- Compare how symmetric and asymmetric encryption work
- Explore how a hash can ensure data integrity and authentication
- Understand the laws that govern the need to secure data
- Discover the practical applications of cryptographic techniques
- Find out how the PKI enables trust
- Get to grips with how data can be secured using a VPN

Packt is searching for authors like you

If you're interested in becoming an author for Packt, please visit `authors.`
`packtpub.com` and apply today. We have worked with thousands of developers and
tech professionals, just like you, to help them share their insight with the global tech
community. You can make a general application, apply for a specific hot topic that we are
recruiting an author for, or submit your own idea.

Share Your Thoughts

Now you've finished *Cryptography Algorithms*, we'd love to hear your thoughts! Scan the
QR code below to go straight to the Amazon review page for this book and share your
feedback or leave a review on the site that you purchased it from.

`https://packt.link/r/1789617138`

Your review is important to us and the tech community and will help us make sure we're
delivering excellent quality content.

www.ingramcontent.com/pod-product-compliance
Lightning Source LLC
Chambersburg PA
CBHW080907220326
41598CB00034B/5499